Culture, Technology, Communication

SUNY series in Computer-Mediated Communication
Teresa M. Harrison and Timothy D. Stephen, Editors

Culture, Technology, Communication

Towards an Intercultural Global Village

EDITED BY
Charles Ess

with Fay Sudweeks

Foreword by
Susan Herring

STATE UNIVERSITY OF NEW YORK PRESS

Published by
State University of New York Press, Albany

Printed in the United States of America

For information, address State University of New York Press,
90 State Street, Suite 700, Albany, NY 12207

Production by Diane Ganeles
Marketing by Patrick Durocher

Library of Congress Cataloging-in-Publication Data

Culture, technology, communication : towards an intercultural global village / Charles Ess, editor, with Fay Sudweeks ; foreword by Susan Herring.
p. cm. — (SUNY series in computer-mediated communication)
Includes bibliographical references and index.
ISBN 0-7914-5015-5 (alk. paper) — ISBN 0-7914-5016-3 (pbk. : alk. paper)
1. Information society—Cross-cultural studies. 2. Information technology—Social aspects—Cross-cultural studies. 3. Intercultural communication. I. Ess, Charles, 1951– II. Sudweeks, Fay. III. Series.

HM851 .C85 2001
303.48'33—dc21 00-061926

10 9 8 7 6 5 4 3 2 1

Contents

Foreword

"The world is getting smaller." This common metaphor is at work in the term "global village," which derives its oxymoronic appeal from the typically small size of a "village" in contrast to the vastness of the "globe." Compared to one hundred years ago, we now have more information about other peoples and cultures, and easier and faster access to that information. Moreover, increased contact has led to the spread—sometimes through imposition, sometimes through voluntary adoption—of Western (especially US) cultural practices. Traditional dress has been replaced by suits in business settings in every country in the world; young people in urban areas everywhere watch films made in Hollywood, listen to rock and roll, play video games, talk on cell phones, wear jeans, drink Coke, eat pizza (or McDonald's hamburgers), speak English, and increasingly, frequent cybercafes. Part of what makes the world seem "smaller" today is that one is more likely to encounter familiar symbols and practices in geographically distant places than was the case one hundred or even fifty years ago.

This trend is facilitated by communication technologies. In the past, highways and railroads enabled information carried by human messengers or in letters to be transported physically from place to place. Later, the invention of the telegraph and the telephone made possible more rapid transmission of messages without people or objects having to be displaced, and radio and television enabled the simultaneous broadcasting of messages to large, geographically dispersed audiences. Most recently, the Internet has introduced interactive, many-to-many communication that transcends both space and time. Today it is possible to disseminate a message widely, inexpensively, almost effortlessly across the globe to anyone who has the technology to receive it, and for others to respond at their convenience using the same technology. Message traffic has proliferated in response to these technological advances, a tribute to human beings' insatiable desire to communicate with one another.

Some people believe that the increased cross-cultural contact facilitated by computer networks will reduce cultural distances, transforming the world into an "electronic global village." Others, noting computer networking's origin in the US, and the continuing predominance of English-language, US-based content on the Internet today, fear that the technology will accelerate cultural homogenization and further consolidate US cultural hegemony on a global scale. As yet, however, there has been little scholarship that evaluates critically the effects of computer networking on the world's cultures. The present volume contributes towards filling this gap.

The volume takes as its point of departure the assumption that the globalization of computer networking is inevitable, and indeed, is already well underway. Undeniably, Internet use is spreading around the world at a rapid rate. As recently as 1996, only 10% of Internet and World Wide Web traffic was in a language other that English. As of this writing, non-English content has risen to 46%, and it is projected to reach 67% by 2005 (Global Reach, 2000). Among the fastest growing languages on-line are Chinese and Spanish, the two languages with the largest numbers of speakers in the world (English has the third largest number of speakers). Internet access is now available even in poor, struggling nations such as Somalia, and to indigenous ethnic minorities in Latin America. In nations which are already "wired," Internet use continues to spread to ethnic minorities, low income groups, and late adopters. For better or for worse, the world appears to be headed for universal Internet access, or something close to it, reminiscent of the spread of television in previous decades.

At the same time, universal access does not guarantee equal power to shape the technology or choose what content it purveys. That power is still overwhelmingly concentrated in the hands of an English-speaking, Western elite, and is not likely to be shaken loose in the near future. Mother-tongue English speakers comprise 5.4% of the earth's population, yet they are overrepresented by a factor of 10 at 54% of Internet users, and will still be overrepresented (by a factor of six) at 33% of Internet users in 2005. Not coincidentally, most Internet and Web content is permeated by Western values of individual freedom (including freedom of expression), religious agnosticism, open sexuality, and free-market capitalism. For cultures that do not share these values—for example, cultures valuing group harmony, religious faith, sexual modesty, and/or economic restraint—the Internet may be perceived as a vehicle of foreign ideology, and resisted to a greater or lesser extent. Moreover, the technology itself—its codes, software, protocols, and interface designs—incorporates an English-language/Western cultural bias that may limit the ability of users from other cultures to maximize its potentials if not translated or redesigned, often at the cost of making it slower or more prone to error. As Yates (1996: 114) puts it, "English-speaking countries may thus always maintain a competitive edge: they have more advanced and more reliable computer software." How effectively individual cultures and subcultures are able to adapt computer network technology to their own values and uses constitutes a major theme of this book.

The book's perspective is both interdisciplinary and cross-cultural. It is interdisciplinary in that the authors bring diverse disciplinary perspectives to bear on the relationship of CMC technology to culture, ranging from philosophy to cultural studies to communication to systems design. It is cross-cultural in that the authors themselves are based in nine countries in North America, Europe, and Asia. The first three articles introduce theoretical concepts and models pertaining to CMC and culture, followed by nine contributions based on ethnographic praxis which describe the current status and use of CMC in Germany, Switzerland, the US, Kuwait, Japan, Korea, India, and Thailand. Most of these are countries about which little scholarly research on Internet use has previously been published; I found these chapters especially informative and thought-provoking.

Among the many timely topics that the essays in this book address, three seem to me to be especially important:

1. *The nature of CMC.* What are the social and psychological effects of computer-mediated communication, and how do they contribute to (or detract from) the potential for an "electronic global village"? Does CMC promote community? Does it support democratic processes?
2. *Technology diffusion.* What factors determine the speed and manner in which CMC technology spreads to and is adopted by (or resisted by) different cultural groups?
3. *System design.* What components of CMC systems are subject to cultural bias? How can culturally-appropriate systems be designed and implemented? Here, "cultural groups" includes gender and ethnic groups within a single nation, as well as the citizens of different nations states.

The answers to these questions are important regardless of whether one considers the globalization of CMC to be desirable or problematic, since in order to bring about positive outcomes from the use of communication technologies in each of these domains, we must first understand how they work in the broadest possible spectrum of cultural contexts.

Still, the question remains: positive outcomes for whom? This book is written in English, by scholars trained in Western academic practices, who by-and-large are optimistic regarding the new technologies and the ultimate effects of their spread. The voices of the poor, the uneducated, the conservative Muslim or Hindu, the nationalistic Frenchman, the Luddite, or even the "average user" are not represented, and thus the overall picture that emerges is neither complete nor culturally unbiased. Nonetheless, much credit is due the editors for broaching this vital and sensitive topic, thereby opening the door to further discussion and debate.

In short, the globalization of the Internet raises intellectual and social challenges concerning cultural bias in CMC, mechanisms of technology diffusion, and barriers to equitable access. As such, it has practical implications for e-commerce, distance education, law, language policy and planning, cultural preservation efforts, politics, and international security, as well as for computer system and software design. Indeed, as the Internet and the World Wide Web continue to spread to ever more remote corners of the world and to diverse subgroups within individual nations, globalization is arguably the single most important issue confronting scholars and users of computer-mediated communication today. The present volume invites us to consider the effects of computer networking from a global perspective, and to evaluate for ourselves whether they are likely to lead to desirable or undesirable outcomes for humankind.

Susan C. Herring

References

Global Reach. 2000. Global Internet statistics. <http://glreach.com/globstats.html>

Yates, Simeon. 1996. "English in Cyberspace." In *Redesigning English: New Texts, New Identities*, eds. S. Goodman and D. Graddol, 106–140. London: Routledge.

Acknowledgments

With the help of an international team of scholars in diverse disciplines, we co-chaired the first international conference on Cultural Attitudes Towards Technology and Communication (CATaC'98), with the goal of bringing together scholars and researchers whose theoretical reflection and research reports "from the field" would shed greater light on how culture shapes distinctive ways of appropriating and using new communication technologies. Some sixty presenters and participants attended, representing eighteen countries. As we had hoped, the conference brought together both highly theoretical reflections and numerous fine-grained reports on diverse cultural attitudes towards communication as well as reports on what happens in the sometime violent, often productive collisions between the new technologies and distinctive cultures.

This volume is one of the outcomes of CATaC'98. Many of the papers collected in this volume were presented at the conference and appeared in the conference proceedings (Ess and Sudweeks, 1998), but have since been reworked, taking into account the discussions and dialogue that were a significant feature of the conference. It is difficult to do justice to the richness of the conference, with regard to individual presentations and especially to the discussions fostered by an unusually collaborative atmosphere. Respected "old hands" and energetic newcomers minimized matters of academic status while maximizing often passionate dialogue among one another as partners in a shared enterprise. Among other things, we hope this volume not only presents some of the best contributions, but also conveys something of the remarkable spirit of dialogue we enjoyed at CATaC'98.

We were fortunate to receive the support of the Science Museum, London, which served as the venue for the conference. The Science Museum was ideal for several reasons. To begin with, it provided us with a conference venue outside the United States, thus helping us offset the tendency for US-based scholarship to dominate the presentations and discussion. In addition, the Science Museum houses a superb exhibit on Charles Babbage's "Difference Engines" and Lady Ada Lovelace's development of programming for these machines, arguably the most significant mechanical and conceptual ancestors of contemporary computers. Indeed, in 1843, Lady Lovelace raised a question broadly thematic of our conference: "Who can foresee the consequences of such an invention?"

CATaC also received significant financial support from the Technology Assessment Program of the Swiss Council of Science, along with important publicity from a variety of scholarly journals and societies:

Communication and Technology Division, International Communication Association

The Communication Technology Policy Section, International Association for Media and Communication Research

Javnost-The Public, Journal of the European Institute for Communication and Culture (Ljubljana, Slovenia)

The Korean Society, publisher of *The U.S.-Korea Review*

Office of Humanities Communication

Philosophy East and West: a Quarterly of Comparative Philosophy, affiliated with the Society for Asian and Comparative Philosophy

University of Sydney, Australia

Drury University, Missouri, USA

Particular thanks go to Simon Joss and Debbie Cahalane for setting the scene at the Science Museum, to Suzanne Tagg who embellished the scene and was an invaluable liaison between the absentee co-chairs and the local players, and to Sara Gwynn for her assistance during the conference. All these people contributed to creating an illusion of an effortless and seamless conference.

We would further like to recognize the Editorial Board members for shaping the conference and for their contribution and constructive reviewing of submitted papers:

Warren Chernaik, Centre for English Studies, UK

Ian Connell, Wolverhampton University, UK

Colin Finney, Imperial College, UK

Jean-Claude Guedon, University of Montreal, Canada

Teri Harrison, Rensselaer Polytechnic Institute, USA

Herbert Hrachovec, University of Vienna, Austria

Ang Peng Hwa, Nanyang Technical University, Singapore

Thomas L. Jacobson, State University of New York, Buffalo, USA

Simon Joss, Imperial College, UK

David Kolb, Bates College, USA

Willard McCarty, Kings College, London, UK

Cliff McKnight, Loughborough University, UK

Sheizaf Rafaeli, Hebrew University of Jerusalem, Israel

Lucienne Rey, Swiss Office of Technology Assessment, Switzerland

Rohan Samarijiva, Ohio State University, USA

Slavko Splichal, University of Ljublijana, Slovenia

We would also like to express our deep appreciation to Ms. Margo Boles (curator of the Kluge-Ruhe Aboriginal Art Collection, University of Virginia) and Mr. Anthony Wallis (Aboriginal Artists Agency, Cammeray, Australia) for their delightful and efficient assistance in acquiring permission to use Dini Tjampitjinpa Campbell's painting.

On a first level, the painting is a conceptual map of connections between important places—typically, waterholes, important geological formations embedded in the religious/philosophical stories of specific peoples, etc.—and thus serves as a powerful visual metaphor for the Web as connecting information centers. Moreover, the painting is an artifact of the oldest continuous human culture on the planet (estimates range between 30,000 to 60,000 years) and is thus most appropriate for a volume examining culture and cultural changes, especially in the face of various forms of what may amount to electronically-mediated cultural imperialism. In particular, in using the dot style, the painting incorporates modifications of Aboriginal art that are designed to *conceal* elements of the map/story that are reserved only for those deemed by tribal elders/knowledge-holders to be worthy of learning the more complex and intricate aspects of the basic map/story. In this way, the painting specifically reflects a cultural change made in response to the European colonization of Australia—and thus visually represents a specific solution to a central question of this volume: in the face of threats to cultural identity through a homogenizing globalization—how may we preserve distinctive cultural identities while also participating in a global mode of communication?

We invite readers to explore more about cultural attitudes towards technology and communication by reading CATaC and related articles now collected in three journal special issues: Sudweeks, F. and C. Ess, eds., *Electronic Journal of Communication / La Revue Electronique de Communication*, 8 (3 & 4: 1998) (see <http://www.cios.org/www/ejc-main.htm>), Ess, C. and F. Sudweeks, eds., *AI and Society*, 13 (1999), and Sudweeks, F. and C. Ess, eds., *Javnost-the Public*, "Global Cultures: Communities, Communication and Transformation," 6 (1999). Readers are also invited to join the discussion group (catac@hhobel.phl.univie.ac.at), and to follow additional CATaC conferences, beginning with the second one held in 2000 at Murdoch University, Perth, Australia (<http://www.it.murdoch.edu.au/~sudweeks/catac00> or <http://www.drury.edu/faculty/ess/catac00>).

Charles Ess and Fay Sudweeks

Introduction: What's Culture Got to Do with It? Cultural Collisions in the Electronic Global Village, Creative Interferences, and the Rise of Culturally-Mediated Computing

Charles Ess

Beyond McLuhan: Interdisciplinary Directions Towards an Intercultural Global Village

In both popular and scholarly literature, the explosive growth of the Internet and the World Wide Web occasions what communication theorist James Carey (1989) identified over a decade ago as a Manichean debate. On the one hand, the "digerati," including such well-known enthusiasts as Nicholas Negroponte (1995) and Bill Gates (1996), promise the realization of Marshall McLuhan's utopian vision of an electronic global village—a theme reflecting earlier, especially postmodernist celebrations of hypertext and computer-mediated communication, as marking out a cultural shift as revolutionary as the printing press, if not the invention of fire (e.g., Lyotard 1984; Bolter 1986, 1991; Landow 1992, 1994). On the other hand, critics see these enthusiastic claims as, at best, resting on questionable myths (Hamelink 1986; Balsamo 1998; Lievrouw 1998) and, at worst, as an electronic utopianism and boosterism (Calabrese 1993; Gaetan 1995; Stoll 1995). Such boosterism, and an unthinking cultural migration into cyberspace, they suggest, may in fact result in less democracy and freedom—and greater exploitation, alienation, and disparities between the haves and the have-nots.[1]

Carey cautions us, however, that this Manichean dilemma is not especially novel. The dilemma reaches back, rather, to the founding documents of the American experience—to the debates between Jefferson and Madison (see the *Federalist Papers*, numbers X and XIV) concerning the role of the new federal government in subsidizing canals and roads. Since democratic polity requires debate and exchange among citizens, it had been argued since Plato that such polities were "naturally" limited—in effect, by the prevailing communication technologies of direct speech and travel by foot or animal. The concern of Jefferson and Madison was how to overcome these natural limits—a necessity if the new republic of thirteen colonies were to be democratic in any meaningful sense. In a conceptual and philosophical maneuver that Carey believes has become definitive of American attitudes regarding technology, Jefferson and Madison turn to communication technologies—in their day, canals and roads—which could overcome the otherwise natural limits to democratic polity.[2]

In this way, Carey suggests that American culture is shaped from the founding of the Republic with a belief that technology, especially communication technologies, can facilitate the spread of democracy and democratic values. Our tendency to debate new technologies in Manichean terms thus falls out of what amounts to a larger cultural assumption that such technologies may overcome otherwise intractable barriers to democratic polity and, should they fail to do so, only the worst anti-democratic possibilities will be realized.

This Manichean debate, moreover, manifests itself on a global scale in the duality identified by political scientist Benjamin Barber as "Jihad vs. McWorld" (1992, 1995). Barber observes that globalization—brought about in part precisely through contemporary technologies which transfer goods and information with ever greater speed and efficiency—tends towards a homogenous "McWorld" in which all significant cultural and linguistic differences are collapsed into a global consumer culture whose *lingua franca* is English and whose primary cultural activity is trade. In the face of this powerful threat to cultural identity, Barber argues, we thus see "Jihad," the rise of local autonomy movements that can become notoriously violent in the name of cultural survival.[3]

If these Manichean dualities represent prevailing presumptions and debates concerning the exponential expansion of computer-mediated communication (CMC) technologies, these oppositions may not be as intractible as they seem. Indeed, we may question these dualities on several levels, beginning with just the point raised by Carey's analysis of this Manichean debate as distinctively American in character.[4] That is, Carey thereby brings to the foreground the role of culture in shaping our discourse and assumptions about communication technologies and their ostensibly crucial role in sustaining the American values of democracy, equality, free speech, etc. But this suggests in turn two central points. First, the assumptions and values shaping our discourse about CMC technologies may be culturally limited: if we explore cultures outside the American orbit, we may find quite different and distinctive assumptions and values. Second, in doing so, we may find alternative ways of understanding the potentials of CMC technologies that allow us to escape, in particular, the Manichean opposition between computer-mediated utopias and dystopias.

The papers gathered here represent precisely an interdisciplinary effort to explore the role culture plays in forming our fundamental beliefs and values—not only with regard to communication and technology, but still more fundamentally towards such basic values as those that cluster about our preferences for democratic polity, individual autonomy, etc. They do so through the lenses of especially three disciplines:

> **philosophy**—as, among other things, an effort to articulate and critically evaluate fundamental assumptions, including the assumptions regarding *values* (ethics and politics), *reality* (as restricted to the material or not), *knowledge* (what counts as legitimate knowledge and how legitimate knowledge(s) may be acquired), and *identity* (including assumptions about human nature, gender, etc.) that define the *worldviews* definitive of diverse cultures;
>
> **cultural studies**—including, but not restricted to, anthropology, sociology, as well as the "sciences of culture" (*Kulturwissenschaften*)[5] supported in European institutions, and so forth; and
>
> **communication theory**—including intercultural communication.

The papers in Part I, "Theoretical Approaches: Postmodernism, Habermas, Luhmann, Hofstede," introduce us to the major theoretical frameworks shaping contemporary analysis and discourse: postmodernism (Jones), Habermas and Luhmann (Becker and Wehner), and Hofstede (Maitland and Bauer). Part II, "Theory/*Praxis*," consists of case studies and research projects from diverse cultural domains that foreground specific cultural values and preferences, and how these interact with CMC technologies developed in the West. These papers document both cultural collisions and creative interferences as Western CMC technologies are taken up in Europe, the Middle East and Asia. Finally, Part III, "Cultural Collisions and Creative Interferences on the (Silk) Road to the Global Village: India and Thailand,"[6] consists of two papers. These echo the patterns of collision and the emergence of new cultural hybrids out of those collisions documented in Part II. But they also provide both suggestions for software localization (Keniston) and a specific model (Hongladarom) for understanding how CMC technologies may be used to catalyze global communication while preserving and enhancing local cultures.

Taken together, these essays demonstrate three key points:

1. While each theory represented here (including postmodernisms, a Habermasian counter to postmodernism, communication theories, and contemporary efforts to predict network diffusion based on identifiable cultural variables (Hofstede/Maitland, Bauer) is partially successful in important ways, no single current theory satisfactorily accounts for or predicts what happens as CMC technologies are taken up in diverse cultural contexts.
2. Culture and gender indeed play a dramatic role in determining how CMC technologies are taken up, whether in the example of listservs and conferencing in an American classroom (Stewart et al.), or in the multiple cultural collisions documented here in the European context (Rey, Hrachovec), the Islamic world (Wheeler), India (Keniston), and the Asian countries of Japan (Heaton), Korea (Yoon, Fouser), and Thailand (Hongladarom).
3. A middle ground between the polarities that otherwise dominate American discourse in particular can, in fact, be theoretically described and implemented in *praxis*. There is an alternative to either Jihad or McWorld, to either postmodern fragmentation or cultural imperialism in the name of putative universals.

Collectively, then, these essays constitute a distinctive conjunction of theory and *praxis*—one that articulates interdisciplinary foundations and practical models for designing and using CMC technologies in ways that avoid the Manichean dualism of Jihad or McWorld, and mark out instead a trajectory towards a genuinely intercultural global village. Especially as these essays illuminate the role of cultural values and communication preferences in the implementation and use of CMC technologies, they first of all uncover the "cosmopolitanism" of popular conceptions of an electronic global village as paradoxically ethnocentric precisely because it ignores the cultural dimensions of both technology and communication. Indeed, like other forms of ethnocentrism, such popular conceptions, especially as fueled by the rapid commercialization of the Net, threaten to further a globalization process that works only by obliterating all cultural distinctiveness. Second, these essays provide the theoretical and practical insights needed to foster an alternative conception of cosmopolitanism: they suggest that what is needed for an intercultural global village in which cultural differences are preserved and enhanced while global communications are also sustained is a new kind of cosmopolitan, one who—precisely through the recognition of the complex interactions documented here between culture, communication, and technology—can engage in both global and local cultures in ways that recognize and respect fundamental cultural values and distinctive communicative preferences.

To see how this is so, I will first provide an overview of each chapter, followed by a summary of some of the insights and additional questions that emerge from these, both individually and collectively. In the last section, I will turn to a fuller description of the sorts of cultural polybrids suggested by these essays, both individually and collectively, as necessary citizens in an intercultural global village.

Overview

Part I. Theoretical Approaches: Postmodernism, Habermas, Luhmann, Hofstede

Steve Jones, in "Understanding Micropolis and Compunity," reviews a number of familiar communication theorists, including Ong and McLuhan, as he develops his own metaphors of path and field to discuss the influence and meaning of Internet messages. In particular, he takes up Carey's distinction between ritual and transportation models of communication to address compunity, which he defines as the merger of computers with communities and our sense of community. This merger, claims Jones, is strained between the traditions and rituals of real life and the kinds of communication as transportation facilitated through CMC. Jones analyzes four areas—privacy, property, protection, and privilege—as central to possible on-line communities.

His analysis both effectively represents the postmodernist approaches that have dominated Anglo-American analysis of hypertext and CMC, and uncovers important ambiguities in the effort to recapture lost community on-line. Such efforts, according to Jones, are only partially successful, and they introduce in their wake new difficulties distinctive to cyberspace. (Such mixed results and ambiguities, we will see, will be characteristic of several analyses and research projects.)

Barbara Becker and Josef Wehner, in "Electronic Networks and Civil Society: Reflections on Structural Changes in the Public Sphere," build on their original presentation at CATaC'98. They begin with a useful overview of a now classic dichotomy. They start with the enthusiasts who see the Internet as inaugurating a communications revolution that will further issue in a radically new form of direct (specifically, libertarian and plebiscite) democracy. The skeptics, by contrast, argue that the Internet is increasingly shaped by new hierarchies and centralized structures, efforts to control and protect information, and a commercialization that threatens to drown out all other activities besides trade. (Sunny Yoon, as we will see, begins with this same dichotomy, including the same warning against the dangers of commercialization.) They draw on theory, including the important debate in contemporary German philosophy between Luhmann and Habermas, as well as empirical research to develop a middle ground between the optimists and the skeptics.

While the optimists see in CMC the promise of radical democracy, Becker and Wehner, echoing especially postmodern analyses of the fragmenting and decentering effects of CMC, note that the kinds of interactive communications that emerge on the Net are precisely those of what amount to special interest groups—relatively small groups of people, often scattered geographically and culturally, who share some minimal set of common interests and abilities, but not necessarily connected (or interested) in any larger, more commonly-shared universe of discourse concerning widely-shared political issues, etc. Indeed, Becker and Wehner note several additional objections to the optimists' dream of radical democracy. Beyond the very real and thorny problems of maldistribution of the economic resources and infrastructure needed to participate in the Net, they take up Bourdieu's notion of cultural capital to observe that not everyone has the level of education, etc., needed to participate meaningfully in on-line exchanges. (Sunny Yoon will also take up Bourdieu, to also stress anti-democratic elements of the Net.) As well, there is the simple problem of noise: "Through networking, more and more participants have a voice; but because of the increasing number of participants, there is less and less time to listen."[7] Nonetheless, Becker and Wehner draw on Habermas's conception of *Teilöffentlichkeiten* ("partial publics," including professional organizations, university clubs, special interest groups, etc.) as loci of discourses that contribute to a larger democratic process in modern societies. Over against the anti-democratic impacts of CMC, they see this Habermasian notion as describing an important component of how CMC technologies may sustain (within limits) a "civil society" as part of a larger democratic process.[8]

Carleen F. Maitland and Johannes M. Bauer, in "National Level Culture and Global Diffusion: The Case of the Internet," start with a careful inventory of the theoretical and practical obstacles to undertaking especially quantitative research into the impact of culture on the diffusion of technology. In the face of these difficulties, Maitland and Bauer first modify and enhance diffusion theory so that it may take up extant quantitative data to explain and predict technology diffusion on a global level. They then move from theory to *praxis* by providing a case study of such analysis as applied to Internet growth. Previous research has tended to focus on matters of economy and infrastructure with relatively little work in the area of culture, in part because earlier work has shown that economic factors are the stronger predictors of technology adoption. In order to test these findings and their own enhancements of earlier diffusion theory, Maitland and Bauer build especially on the work of Hofstede and Herbig to include three cultural factors in their study: uncertainty avoidance, gender equality, and English language ability.

Their extensive statistical study draws on a considerable range of data sources, as available for 185 countries during the time period between 1991 and 1997. In examining Internet growth between countries, they find that cultural variables are less significant in explaining adoption than economic or infrastructure variables: of these, teledensity, international call cost, and school enrollment emerge as the strongest predictors, the last finding supporting the importance of education in development. For that, the cultural factor of English language ability also plays a significant role. In analyzing growth within countries, their data likewise uncovers a comparatively stronger role for economic factors—in this case, the number of PCs per capita. But cultural factors—namely, uncertainty avoidance and gender empowerment—also play a significant role.

Maitland and Bauer's work is significant because it refines diffusion theory so as to more adequately take into account specifically cultural factors, and as their analysis demonstrates the importance of cultural factors: simply, by including cultural factors along with economic and infrastructure dimensions, their models enjoy an increased predictive power. And, especially for our purposes, their work is important as it provides an empirical basis that demonstrates the impact of important cultural variables on technology diffusion. Finally, their quantitative approach, as confirming the importance of English language ability, meshes well with Becker and Wehner, as well as Yoon, all of whom take up Bordieu's notion of cultural capital (which includes language ability) as a necessary element of cultural analysis; this finding is further consistent with Keniston's observations regarding the role of English as a passport to computing—and thus to power and prestige—in India.

Part II. Theory / Praxis

a. The European Context

Herbert Hrachovec, in "New Kids on the Net: Deutschsprachige Philosophie elektronisch," documents several experiments with conducting philosophy on-line in the German-speaking world, illustrating "the force and limits of attempts to install a computer-mediated space of Reason." Hrachovec is critical of too closely identifying at least the current realities of hypertext with such standard postmodernist theorists as Barthes and Derrida (an identification made most effectively and prominently by George Landow). In particular, it may not be accidental that "electronic philosophy" is very much at the margins of German academic life: "some features of the new discursive forms are incompatible with the current educational system." Hrachovec's study of the contrasts between the "microcultures" (my term) of traditional academia and on-line discourse may point to similar contrasts in larger contexts.

Lucienne Rey, in "Cultural Attitudes toward Technology and Communication: A Study in the 'Multi-cultural' Environment of Switzerland," examines the political differences between the four major linguistic groups of Switzerland—German, French, Italian, and Romansch—and then seeks to determine whether these ethnic/linguistic differences also correlate with different attitudes towards technology. In point of fact, her findings suggest that the German-speaking part of Switzerland, the most politically and economically dominant component of the country, is at the same time the most conservative in the sense that German-speaking Swiss show less openness to and interest in the new communications technologies than their Latin compatriots. Rey helpfully suggests that this cultural attitude may have two roots. First, she notes that German scepticism towards progress through technology is rooted in the German Romantic tradition, as this tradition reacts against the Enlightenment and the early stages of mechanization as brought about by the Industrial Revolution. Two, she observes a contrast between the playfulness of the Swiss-French and the seriousness of the Swiss-Germans. Given the playful dimensions of interactions on the Net and the Web, she hypothesizes, they are likely to be more attractive to the French than the Germans.

b. Gender/Women in Islam

Contrary to the common presumption that CMC technologies bring about greater openness and democratization, Concetta Stewart, Stella F. Shields and Nandini Sen, in "Diversity in On-Line Discussions: A Study of Cultural and Gender Differences in Listservs," begin with the recognition that women and minorities have historically enjoyed less access to these technologies. To better understand this exclusion, they explore in their own study how two sorts of differences in communication style appear in listservs: cultural differences first articulated by Hall between high- and low-context cultures (and supplemented here by Ting-Toomey's Face-Negotiation Theory); and gender-related differences, documented by Tannen and Herring. Their rich overview of earlier research into gender and cultural variables (including those delineated by Hofstede) in cross-cultural communication theory demonstrates that while there is a significant body of research in intercultural communication, cross-cultural communication in CMC environments has been relatively ignored until now. Their study of an in-class listserv, intended to further free and open communication among a considerable diversity of students, strikingly confirms that gender and culture profoundly limit how far conversation on listservs may be said to be open and democratic.

Just as elsewhere, the Internet and the Web are of compelling interest in the various countries and cultures centrally shaped by Islamic values and traditions. And this is despite a possible mismatch between the "high content/low context" communication preferences which have shaped the Western development of these technologies versus the "high context/low content" character of communication in Arabic societies.[9] Deborah Wheeler, in "New Technologies, Old Culture: A Look at Women, Gender, and the Internet in Kuwait," takes up the familiar promise claimed by Western proponents of CMC technologies—that they will promote democracy, prosperity, and equality, including gender equality—and tests this promise against a careful ethnographic study of Kuwaiti women and their use of the Internet. Her case study is valuable first of all as it sheds light on a little researched but critically important series of intersections: Islam and sharply-defined gender roles vis-à-vis a communication technology hailed by Western feminists for its promise of expanding gender equality. In addition, Kuwait is especially instructive insofar as it enjoys one of the highest per capita incomes in the world. These and other characteristics mean that if there is resistance to new CMC technologies, such resistance is not obviously the result of infrastructure deficits, an entrenched anti-technology culture, or extreme patriarchal structures.

Wheeler's analysis of how far the Internet and the Web serve the cause of gender equality shows decidedly mixed results. On the one hand, her interviews with younger women support the notion that these new technologies do have a liberating impact. For example, they allow women to converse "unescorted" with men in chat rooms, and to meet and choose mates on their own (rather than agree to the cultural norm of arranged marriages). At the same time, however, she finds that the powerful restrictions against women speaking openly in Kuwait are directly mirrored in differences between women's and men's characteristic use of CMC technologies. As she observes, "The advent of new fora for communication does not automatically liberate communicators from the cultural vestiges which make every region particular and which hold society together." While Wheeler concludes on a hopeful note, she reminds us nonetheless that activism is always local and thus shaped by specific institutional and cultural imperatives.

c. East-West/East

Contrary to the view that technologies are value and culturally neutral, in "Preserving Communication Context: Virtual Workspace and Interpersonal Space in Japanese CSCW," Lorna Heaton presents two case studies to show how cultural values and communication styles specific to Japan are incorporated in the design of computer-supported cooperative work (CSCW) systems. She does so from a social constructivist view, one that further suggests that technologies can be "read" as texts, and drawing specifically on Bijker and Law's notion of technological frame to explain how Japanese designers invoke elements of Japanese culture in justifying technical decisions. Heaton highlights the importance of nonverbal cues and the direction of gaze in Japanese culture as an example of Hall's "high context/low content" category of cultural communication style, in contrast with Western preferences for direct eye contact and "low context/high content" forms of communication. She also notes in her conclusion the Japanese interest in pen-based computing, speech synthesis, virtual reality interfaces, etc., as resulting not only from the physical difficulties of using a Roman keyboard to input Japanese, but also the larger cultural preference for high context in communication.

Sunny Yoon, in "Internet Discourse and the *Habitus* of Korea's New Generation," counters the familiar portrayal of the Internet as a medium that will engender greater democracy, especially in the form of an electronic "public sphere" (a requirement for democracy, according to Habermas). She notes the ways in which the Net, especially as it becomes ever more commercialized, may work rather as a controlling mechanism for capital and power. Here, she takes up Foucault once again (see Yoon 1996), along with Bourdieu's notion of *habitus*, as frameworks for analyzing power as manifested in the workings and impacts of the Net.

In contrast with other postmodernist concepts, the notion of *habitus* emphasizes individual will power and choice; these manifest themselves in individuals' everyday practices which in turn, in an "orchestra effect," build up the larger society and history in which individuals participate. Such *habitus* clearly influences individual choice, but not in fully deterministic ways.[10] Moreover, Bourdieu sees "cultural capital" (including symbolic and institutional power—most prominently, language and education) as creating the *meconnaissance* ("misconsciousness") of the majority, a kind of false consciousness which legitimates existing authorities.

Yoon first presents her careful quantitative study of Korean newspaper reports on the Internet and on-line activities. Her analysis makes clear that Korean journalism fails to encourage the use of the Internet as a medium of participatory communication. Rather, Korean reporting contributes to the commercialization of the Internet and thereby, some argue, unequal access to and distribution of information resources. Yoon then turns to a series of ethnographic interviews with young Koreans ("Gen-Xers"). While she is careful to recognize that the results of her small sample cannot be generalized, her interviews demonstrate that the Internet exercises symbolic or positive power—including symbolic violence in Bourdieu's sense—as it shapes educational rules and linguistic habits. In particular, Korean students accept the on-line dominance and importance of English without question. Language thereby becomes a cultural capital that exercises ". . . symbolic power over the cultural have-nots in the virtual world system," a cultural capital that induces a "voluntary subjugation." At the same time, however, Yoon documents how individuals take up the Internet, not because of its promise of greater equality and democracy, or even utility, but, rather to the contrary, because it increases their status, and thereby their distance from and power over others. As well, the comparative expertise of young people gives them considerable power over their elders because teachers, principals, and parents rely more and more on the younger generation to help them learn how to use computers, design institutional documents and web pages, etc. Contrary to the presumption that the Internet only democratizes, Yoon demonstrates that the Internet, by shaping *habitus* in these ways, can lead either to resistance or subjugation, to democratic communication, or (cultural) capitalist dominance. Consequently, she argues, we must better understand the concrete processes of how the Internet functions as the *habitus* of people in their everyday lives before attempting to decide which of these two directions the Internet might take us.

Robert Fouser, in "Culture, Computer Literacy, and the Media in Creating Public Attitudes toward CMC in Japan and Korea," brings together a wide range of information (a review of web sites vis-à-vis print media, attitudinal survey data, comparative studies of GNP and CMC infrastructure, recent scholarship, and personal interviews) to develop a clear picture of the striking differences between Japan and Korea with regard to attitudes towards and utilization of new communications technologies, including CMC technologies. It may come as a surprise to Westerners to learn that while Japan is materially wealthier than Korea, and perhaps better known in the West for its prowess in developing and marketing new technologies, Koreans show a greater interest in and usage of CMC technologies than the Japanese. Fouser reviews two theories that might explain these differences. The first is a "culture" theory which focuses on a shared set of values and attitudes; the second is a "computer literacy" theory that looks instead to the pragmatic elements of cost, and ease of use. For example, Korean, as a language which, like English, uses an alphabet system rather than the highly complex character systems of Japanese and Chinese, is much easier to enter through a keyboard than Japanese or Chinese. Fouser finds that the notion of "culture" is too broad to account for a Japanese lack of enthusiasm for CMC in particular, over against their more positive attitudes towards other new technologies (including mobile phones). Instead, he argues that more pragmatic elements, including political leadership in encouraging the use of new technologies, are better predictors of technology diffusion.

First of all, then, Fouser's work—especially as read together with Yoon—helps us develop a more nuanced understanding of how CMC technologies are taken up in two distinctive Asian societies. Secondly, his work illustrates the limits of cultural approaches to questions of technology diffusion and helpfully demonstrates that such cultural approaches must be complemented with pragmatic considerations of political leadership, etc. In this second direction, his work should be taken together with the several other contributions gathered here, including Maitland and Bauer's quantitative analysis of culture, that both individually and collectively help us better understand the difficulties of developing meaningful definitions of "culture"—and the necessity of complementing even the best definitions with additional conceptual frameworks if we are to develop a more complete understanding of the interactions between technology and culture.

Part III. Cultural Collisions and Creative Interferences on the (Silk) Road to the Global Village: India and Thailand

Some of the first indications that Western-based CMC technologies did indeed implicate culturally-distinctive values that would clash with the values and preferences of other cultures were documented in Asia.[11] Two final studies in this collection—the first on localized software in India, the second on an "electronic Thai coffee house"—document how local cultural values indeed collide with the values apparently shaping Western CMC technologies.

But these two chapters further demonstrate that cultural collisions [and with them, the danger of imperialism and "cultural steamrolling" (Steve Jones 1998)] are not the whole story. Rather, Kenneth Keniston argues for ways to overcome the otherwise daunting obstacles to "localizing" software. Yoon and Fouser amply demonstrate the power of English as the *lingua franca* of the Web: localization seeks to counter this power on a first level, as Keniston explains. Such localization, however, requires not only translation of documentation and commands into another language: such transformation also extends to interface design (including icons, use of color and other symbols which vary—sometimes dramatically—in their meaning in diverse cultures), and to the underlying machine codes (such as ASCII and Unicode) which must be universal if computers and networks are to successfully communicate with one another (cf. Pargman 1999). On all these levels, the current standards are predominantly the products of Western, English-speaking computer designers and software writers. Keniston suggests ways of overcoming these obstacles in the Indian case and thereby points to how Indian efforts to localize software may be paradigms for other cultures that seek to be members of the global village while preserving local languages and cultural values. Soraj Hongladarom's account of Thai discussion groups provides a powerful example of Keniston's hope for such dual citizenship (i.e., global/local). At the same time, Hongladarom connects this dual citizenship with significant theory: he makes use of Michael Walzer's analysis of "thick" and "thin" cultures to suggest what might indeed be a model for an electronic global village which both facilitates the global and preserves the local.

Kenneth Keniston, in "Language, Power, and Software," takes up the role of language in the development and diffusion of computer technologies, specifically with a view towards how the predominant language of computing—English—reinforces current distribution patterns of "power, wealth, privilege, and access to desired resources." The problem of such linguistic imperialism (my term) is especially clear in efforts to localize software—transforming software to make it useable by those outside the cultural domains defined by English. In addition, English-only access to computing technologies also exacerbates the larger global tension identified by Barber in terms of "Jihad vs. McWorld." As Barber makes clear, finding a middle ground between these two poles is crucial for the survival of some form of participatory democracy: Keniston emphasizes the point that such a middle ground is crucial for the survival of local cultures and languages.

India is an especially compelling case study for examining these concerns. India is the world's largest democracy, a nation that further encompasses a breathtaking diversity of languages, including eighteen official languages and some three hundred unofficial spoken languages (Herring 1999b). Where English is the privileged route to power, less than 5% of these populations speak

English. But there is almost no readily available vernacular software in India.

Keniston identifies a number of fundamental obstacles to localization, including local cultural factors that weigh against localization—factors resulting from both an indigenous religious tradition and British colonialism. On the one hand, the Brahmanic emphasis on higher levels of spirituality, thought, and action, in contrast with the earthly and material, means that writing localized software programs "for the masses" seems less important than other pursuits. On the other hand, the success of British colonialism has meant precisely that English is the prestige language in India. Hence, to program in English (e.g., for export) is laudable, while programming in an indigenous language is to run contrary to the cosmopolitan trajectory affiliated with English, and to run the risk of seeming the ally of "fundamentalism" and the tribal (Jihad in Barber's sense). And since localized software would provide access to computer technologies—and thereby, to the power, wealth, and prestige such technologies are affiliated with—for those traditionally excluded from elite status (outcastes, tribals, etc.), such software may be seen as a direct threat to the privileges enjoyed by those who would write the localized code.

Despite such obstacles, Keniston closes with a series of suggestions intended to encourage the localization of software needed if the new technologies are to help close, rather than widen, the gap between the haves and the have-nots—and if the new technologies are to help enhance cultural diversity rather than eliminate it. As Keniston notes, these difficulties are especially acute in South Asia because of its distinctive fusion of power and language. At the same time, however, successful solutions to the localization problem in South Asia are likely to serve as models for preserving democracy and cultural diversity on a more global scale as well.

How we avoid Manichean choices is the lesson suggested by Soraj Hongladarom, in his "Global Culture, Local Cultures, and the Internet: The Thai Example." Hongladarom examines two threads of discussion developed in a Thai Usenet newsgroup, one dealing with critiques of the Thai political system and the other with the question of whether Thai should be a language, perhaps the only language, used on the newsgroup. In contrast with concerns that CMC technologies will erase local cultures and issue in a monolithic global culture, Hongladarom argues that the Internet facilitates two different kinds of communication: (1) communication that helps reinforce local cultural identity and community (in part, as this communication fulfills what Carey calls the "ritual function", i.e. strengthening community ties); and (2) communication that creates an "umbrella cosmopolitan culture" required for communication between people from different cultures. Hongladarom further suggests that we distinguish between a Western culture which endorses human rights, individualism, egalitarianism and other values of a liberal democratic culture (a "thick" culture in Walzer's terms), and the cosmopolitan culture of the Internet as neutral (a "thin" culture).[12] The Thai experience suggests that the Internet does not force the importation of Western cultural values. Instead, Thai users are free to take up such issues and values if they wish, and they can do so while at the same time preserving their cultural identity.[13]

A First Philosophical Response: Whither the Electronic Global Village?

These essays demonstrate the importance of cultural attitudes in shaping the implementation and use of CMC technologies, whether those technologies are introduced within distinct but still Western cultures (Hrachovec and Rey) or in the diverse cultures of Asia and the Middle East. First of all, these chapters directly call into question the characteristically American confidence in communication technologies as making possible democratic discourse and equality, especially when confronted with the radical linguistic and cultural diversities of India (Keniston) and the deeply entrenched gender roles of Kuwaiti society (Wheeler).

These essays likewise counter the Manichean dualities of American discourse, whether in terms of cyber-utopias (including McLuhan's global village) versus cyber-dystopias, or Barber's double dystopia of Jihad versus McWorld. Rather, Heaton's account of Japanese redesign of CSCW systems and Hongladarom's experience and model of a "thin" Internet culture coupled with "thick" local cultures (especially as facilitated by localized software, as Keniston recommends) demonstrate first of all that these technologies indeed embed and abet specific cultural communication preferences (such as for high content/low context vs. low content/ high context) and values (democratic polity, equality, etc.). However, they are not unstoppable forces. On the contrary, they can be localized and reshaped—and stripped, if necessary—of the cultural values and preferences they convey.

In philosophical terms, the hopes of computer-mediated heaven and fears of cyber-hells rest on a view called technological determinism. Such a view sees technology and whatever effects follow in its wake as possessing their own autonomous power, one that cannot be resisted or turned by individual or collective decisions.[14] The hope of proponents is that the introduction of CMC technologies will inevitably change cultural values for their own good. These technologies will convey and reinforce preferences for, say, free speech and individualism, particularly in the case of the Internet and the Web, as centralized control of information conveyed through these technologies is very difficult.[15] In the inverse dystopian image, captured powerfully in the images of the Borg in *Star Trek*, technology is likewise an unstoppable force; once infected by the Borg implants, all humanity (meaning specifically such qualities as individuality, compassion, and choice) is lost as one becomes seamlessly integrated into the single-minded machinery of the Collective. Such science-fiction portrayals nicely capture the real-world fears of those who see CMC technologies as central engines in the global but homogenous McWorld that will override and eliminate local choice and distinctive cultural values.

But consonant with philosophical critiques,[16] such (hard) technological determinism is clearly belied by these studies, beginning with Jones' analysis of the limits of any on-line community. Such a "compunity," to use his term, is more likely to emerge as a micropolis rather than the cosmopolis of a single global culture. And as Yoon makes clear in her analysis, the *habitus* of cultural practices and attitudes surrounding computing exercises a kind of cultural power that can be both shaped and resisted by individuals. This suggests that both individuals and countries can make choices regarding how the implementation of CMC technologies will shape their political and cultural futures. Most powerfully, Hongladarom's example of "thin" Internet culture/"thick" local cultures stands as a concrete alternative to such Manichean dualisms—one instantiated in *praxis* in the Thai case.[17] Negatively, these analyses and examples thus contradict the assumption of (hard) technological determinism and with it, the Manichean dualities that rest upon this assumption. Positively, they identify middle grounds between a McWorld that steamrolls local cultures and the Jihad that such imperialism and homogenization may evoke.[18]

From Philosophy to Interdisciplinary Dialogue: Cultural Attitudes towards Technology and Communication

Technological determinism is not the only assumption underlying the prevailing icons of what Keniston identifies as the Anglo-Saxon discussion of CMC technologies. As we saw in the opening paragraphs, McLuhan's global village and its attendant Manichean polarities further implicate what now appears to be an especially American presumption that communication technologies are crucial for the survival and expansion of democracy and individual freedom. Moreover, especially from a philosophical approach, a range of additional presumptions can be seen to underlie the optimistic vision of an electronic global village; presumptions, moreover, which are quickly entangled in paradox and contradiction.

To begin with, such a vision is clearly cosmopolitan in its assumptions and intentions. As traced back to as far as the Stoic philosophers of the Greco-Roman world, this vision rests on an optimistic conception of a shared (and essentially rational) humanity, one capable of becoming the *cosmo-politan*—the citizen of the world—not simply the citizen of a given country and culture. This cosmopolitan trajectory is consciously developed to counter the ethnocentrism characteristic of prevailing cultures (i.e., the belief that one's own language/culture/worldview are the only "right" ones, and those who adhere to differing languages/cultures/worldviews are simply wrong, inferior, etc.).

In light of the role of culture in shaping fundamental assumptions, however, we can raise this question: Is this ostensibly cosmopolitan image, as it intends to overcome the ethnocentrism of particular cultures (as based on specific traditions, habits, prejudices, etc.) with a universally-shared humanity, itself ethnocentric as it rests upon culturally-limited assumptions, beginning with the characteristically American belief in communication technology as central to the spread of democratic polity? In other words, is this cosmopolitan vision itself a form of "cyber-centrism," an ethnocentrism in its own right that runs in tension with its cosmopolitan intentions?

Similarly, the conception of an electronic global village seems to presume that the tools of CMC—the computer codes, interfaces, etc.—are culturally neutral, i.e., they allow perfectly transparent communication between members of all cultures, without giving preference to the distinctive values and communication preferences of any single culture. Philosophers denote this presumption as "technological instrumentalism." At the same time, however, we have already seen that the electronic global village also presumes a technological determinism, the view that CMC technologies are not culturally neutral, but in fact embed, convey, and reinforce specific values such as individualism, free speech, etc. Thus, the McLuhanesque vision of an electronic global village appears to rest on two mutually contradictory assumptions: if technology determines its users along specific value sets, it is clearly not value-neutral, and if it is value-neutral, then it clearly cannot determine its users along specific value sets. Moreover, both philosophical assumptions—technological instrumentalism and (hard) technological determinism—are called into serious question on both theoretical and practical grounds in the chapters collected here and in the larger literature.[19]

Since Aristotle, philosophers have recognized that theory must be tested and engaged in *praxis* (cf. *Nichomachean Ethics,* esp. 1179a35–1179b3). (Admittedly, philosophers have not always practiced this recognition!) To determine more carefully the fundamental assumptions underlying the prevailing conceptions of an electronic global village—including their potential paradoxes and contradictions—thus requires nothing less than an inquiry on a global scale into what happens in *praxis* as CMC technologies are taken up in diverse cultures. Such an inquiry, moreover, is by no means of interest only to philosophers. Rather, it requires and intersects directly with the full range of methodologies, approaches, and insights of multiple disciplines, beginning with communication theory and cultural studies. And of course, no single scholar or researcher can hope to undertake such an inquiry as a solitary exercise. This global inquiry simply requires an interdisciplinary dialogue of global scope.

The first conference on Cultural Attitudes towards Technology and Communication (CATaC'98) was devoted to just such an interdisciplinary global dialogue. As noted above, the papers collected here—most originally presented at CATaC'98—represent some of the best contributions. At this point, it may be helpful to note the strengths and limits of CATaC'98, in order to develop a more complete understanding of the larger context of these chapters, including the trajectories for future research they and CATaC'98 limn out.

Cultural Limitations

On the one hand, CATaC'98 achieved an exceptional scope in terms of the cultural domains represented by participants and presenters: studies included North/South, East/West, Industrialized/Industrializing, and Colonial/Indigenous countries/peoples.[20]

But there were also striking absences: China, France and the Francophone countries (except Switzerland) and Arabic/Islamic countries were not represented.[21] For that, in this volume, Deborah Wheeler's study of Internet usage in Kuwait provides important insights into network diffusion in the Islamic world, especially with a view towards the role of gender.

Theoretical Limitation: Religion

"Religion" is ordinarily recognized as a major source (either directly or indirectly) of the worldview of perhaps all people. Nonetheless, religion is striking for its absence in these papers—again, with the exception of Deborah Wheeler's study of women in Kuwait.

This absence raises several questions. American academic culture, for example, seems uniformly hostile to raising questions of religion, at least outside of religious studies and some sociology circles. This disciplined silence, no doubt, has several roots, ranging from the influence of positivism (which simply discarded all religious claims as nonsense while re-explaining them in materialist terms) in the academy to a characteristically American notion that "religion" is a matter of private concern only, one not to be brought up in polite society.

Such silence is a sensible strategy in the face of the power of religious issues to (literally) explode the fabric of civil society, as they have done throughout much of Western history, including early American colonial experience, contemporary UK experience, etc. But it seems clear (as Wheeler's chapter demonstrates) that any adequate account of "culture" and CMC must squarely face the religiously-shaped components of culture and worldview, or demonstrate that religion is fully reducible to the components of culture identified by Hofstede, Hall, etc.

Theoretical Issues and Questions: Culture and Worldview; Postmodernism, Habermas, and Hermeneutics

As noted in the opening paragraphs, no single theory yet adequate accounts for all the complex interactions between culture, technology, and communication. First of all, as Rey points out, one of the central conceptual challenges for any theory—and thereby, any empirical study—is to provide a satisfactory account of what "culture" means. By operationalizing her definition of culture in terms of linguistic boundaries, Rey is able to provide her most intriguing empirical analysis of the contrasts between German- and Latin-speaking Swiss. Heaton's use of Hofstede and others also shows the power of developing operational definitions (see also Smith et al. 1996). And both Heaton and Yoon add to this operational approach in part as they take up Bourdieu's notion of *habitus*. Maitland and Bauer also provide a helpful overview of possible definitions, beginning with Clifford Geertz's widely used account; they further note that culture includes norms and values that are not necessarily isomorphic with linguistic and national boundaries and thereby indicate the limits of operational definitions that identify "culture" solely with language.

In doing this, Maitland and Bauer further make explicit one of the central intersections between communication theory, cultural studies, and philosophy: if culture explicitly includes norms and values, it thereby involves what philosophers and anthropologists study as "worldview." Lacroix and Tremblay (1997) point out that as the term "culture" refers to norms and values, it thereby refers to the non-material, and thus to the province of philosophy, including epistemology.[22] Since Aristotle, philosophers have recognized that the non-material character of values and norms means in part that they can be known with less precision and agreement on their meaning than, in Aristotle's example, the axioms of mathematics (*Nichomachean Ethics* 1094b13–27). To develop a satisfactory account of what "culture" means, then, seems to require just the interdisciplinary efforts of philosophers, cultural scientists, and communication theorists (among others): to develop such an account remains a central theoretical challenge.[23]

But in addition, while no single theory may be complete, the diverse range of theories invoked in this work allow for one theory to complement the deficits of others. For example, at CATaC'98, Cameron Richards echoed a common critique of the postmodern approaches otherwise fruitfully represented here by Jones, Becker and Wehner, and Yoon. Richards pointed out that postmodern frames, while useful, cannot justify any normative judgment that distinguishes between the use and abuse of CMC technologies, i.e., between precisely the utopian futures (because more democratic, egalitarian, etc.) they characteristically endorse and the dystopian possibilities they shun (because more totalitarian, hierarchical, etc.). This critique meshes with more broadly philosophical critiques of postmodernism as relativistic and thus incapable of grounding its endorsement of democracy over fascism, of equality over privilege, etc.[24] To offset this deficit, Richards (1998) turns to Paul Ricoeur's hermeneutical approach as providing ways of more coherently justifying our preferences for the utopian possibilities of CMC technologies. Similarly, in this volume, Barbara Becker and Josef Wehner take up Habermas's notion of *Teilöffentlichkeiten* (partial publics) as a way of countering postmodern emphases of fragmentation, decentering, chaos, etc. In this way, both contributions present a model of theoretical complementarity or pluralism that attempts to hold together more than one theoretical approach, using the strengths of one to complement the limits of another. Such pluralism is manifest more broadly in just the interdisciplinary dialogues represented here between philosophy, communication theory, and cultural studies. This pluralism and dialogue, most broadly, are the theoretical counterparts to the models suggested especially by Keniston and Hongladarom; to repeat, they collectively argue for a dual citizenship in a "thin" but global Internet culture and in one (or more) of the great diversity of local "thick" cultures ideally sustained in an intercultural global village. But while these sketches may serve to suggest the initial outlines of a more complete theory encompassing culture, technology, and communication, work in this area appears to only have just begun.[25]

Moreover, Richards noted the postmodernist tendency to sharply distinguish between real and "virtual" identities, so as to claim that cyberspace represents genuinely radical and revolutionary change in our current conceptions of identity, community, etc. In discussion at CATaC'98, Richards suggested that, nonetheless, "the individual voices of cyberspace are somehow still embodied, and thus still connected to physical and thus cultural realities." Richards' analysis on this point can be fruitfully compared with the work of Susan Herring, who has now extensively documented gender differences in the ostensibly "gender blind" spaces of CMC (Herring 1999a).

Theoretical Issues and Questions: Embodiment and Gender

Steve Jones, in his summary comments on CATaC'98, reiterated the importance of more attention to the issues of embodiment and gender. Gender is addressed, for example, when Maitland and Bauer note that network diffusion is positively affected by Hofstede's cultural dimensions of gender equality—and, in this volume, in Wheeler's account of women in Kuwait. While there is no shortage of research on gender differences and culture (e.g., Smith et al, 1997), more attention is needed to the construction of gender within given societies and how diverse expectations concerning gender interact with CMC technologies.

Indeed, the focus on embodiment and a correlative recognition that (most) human beings cannot jump out of their embodied/gendered cultural identities may work in support of Hongladarom's model of "thin" but global Internet culture coupled with "thick" local cultures. Such a model stands as a middle ground between cultural conservativism and isolationism (Jihad) versus radical and revolutionary cultural transformation. In doing so, it further points to the central importance of embodiment in our understanding human beings as participants in and shapers of cultural traditions. By contrast, the enthusiasts' emphasis on the radical transformations to be brought about through the rise of cyberspace often rest on a kind of cyber-gnosticism—a dualistic (indeed, Manichean!) opposition between body (as implicated in the web of real-life relationships, communities, etc.) and mind (as capable of full self-expression in cyberspace). Such cyber-gnosticism is not only apparent in the (early) cyborg feminism of Donna Haraway, who endorsed escape from real-life gender discrimination into the ostensibly gender-blind and gender-equal domain of cyberspace; it is further at work in the libertarian rejection of real-life political communities, including their limits on free speech, by such spokesmen for the American Internet culture as John Perry Barlow, a co-founder of the Electronic Frontier Foundation.[26] It may not be accidental that such Manichean/ Gnostic contempt for the body can be found alongside the Manichean dualities emphasizing that salvation can only be found by escaping the body in cyberspace—especially given the prevailing context of an American discourse defined largely by just such Manichean dualism. By turning instead to a recognition of the role of embodiment as intertwined with the ways in which culture has us communicate and interact with technology, we may develop theoretical understandings of our connection with and freedom from body and culture more consonant with the middle course of both preserving and moving beyond our local cultures.[27]

Preliminary Conclusions: Cultural Collisions, Cultural Hybrids, and Intellectual Mutts—Considerations for Becoming Citizens in the Electronic Global Village

Physicists seek to infer the properties of otherwise hidden particles by carefully examining what happens when these particles collide at high energies. Encountering a culture distinct from one's own—a culture whose patterns of life, including language, customs, and values, may differ radically from those defining the world one has previously inhabited—involves analogous collisions. Collisions occur between underlying assumptions, including basic ethical and political values and communicative styles that make up the worldview characteristic of each culture. "Culture shock" is the name we give this experience. In part, the shock involves precisely the realization that what one has presumed, perhaps for all of one's life, to be universally human ways of talking, believing, valuing, are instead limited. Other peoples, other cultures, do and believe differently, and usually seem to thrive in doing so. As the properties of invisible particles may be inferred from the traces and debris of their collisions, so culture shock allows us to uncover the usually fundamental but tacit assumptions of our and other cultures, as it forces us to make explicit the manifold presumptions of colliding worldviews.

Such collisions, in fact, are not the whole story. When we, as newcomers, seek to become oriented in a new place, we sort through what is radically different and what seems shared ("Everyone cooks with water," the Swiss say). Gradually, we may find that what initially seemed alien is not so strange. Indeed, many of us often find that some beliefs and habits of other cultures make more "sense" than our own, and we seek to sustain those ways of being when we return to our own places (making us seem very odd ducks indeed to those neighbors who have not had the privilege of living elsewhere). In most cases, we do not reject all of our original beliefs and values ("going native," as it is said). We become, instead, what is variously described as multicultural (Adler 1977) "intercultural" (Gudykunst and Kim 1997), or "Third Culture Persons" (Finn-Jordan 1998)—that is, multilingual cultural hybrids, able to travel, speak, and live (in varying degrees of facility) in more than one cultural domain.[28]

This process of making explicit and sifting through the fundamental elements of diverse worldviews, and constructing new hybrid views and ways of being, is one focus of intercultural communication, and can be aided by cultural studies (see Samovar and Porter 1988, esp. Part 4; Bennett 1998). At the same time, this process engages us in several of the distinctive tasks of philosophy: identifying both our and others' most fundamental assumptions concerning what is real, how do we know, who are we as human beings, what ought we to value and disvalue, etc.; critically evaluating differing beliefs; and attempting to determine for ourselves just what we may hold to be true in a new synthesis of views. In Plato's well-known allegory of the cave, this process involves precisely leaving the world of one's everyday experience—one's own city or culture. But once a more complete understanding of the *cosmos* is achieved, the philosopher returns to the place where she started, seeking to integrate her new understanding with the familiar beliefs and habits of her cohorts, and to encourage her cohorts likewise to achieve a more complete understanding of what lies beyond the boundaries of their own *ethnos*, beyond their ordinary experience of the everyday. Similarly, in the religious stories of many cultures, this task—the discovery of realities beyond ordinary experience, and the integration of these insights and noumenous powers into the everyday—is central to the process of growing up.[29]

On the academic level—and in a more homely metaphor—the scholars and researchers who presented and discussed at CATaC'98 described themselves as intellectual mutts, as hybrids and crossbreeds who could not be categorized within a single discipline. And so the essays collected here likewise cross boundaries. As they document cultural collisions and collusion from interdisciplinary and intercultural perspectives, they may contribute to our readers' own discovery of new cultural and communicative views and beliefs and thereby contribute to their own boundary crossings (academic and beyond) and resulting constructions of more complete, multicultural worldviews.

Indeed, becoming such multicultural persons in these ways is not simply a project of individual significance, reserved only for the few (as Plato's allegory suggests). Rather, these essays argue in at least two ways that our becoming multicultural is a necessary component of an electronic global village that aims towards an intercultural synthesis of the global and the local. Most apparently: Keniston's model of dual citizenship in what Hongladarom describes as a thin global culture and thick local cultures requires that such citizens themselves become cultural hybrids—precisely the multicultural persons who can integrate and live in multiple worlds. Secondly, and at a still more philosophical level: these essays undermine both technological instrumentalism and (hard) technological determinism. This means especially that an electronic global village marked by specific human values—including respect for cultural diversity—will not emerge automatically on its own as an inevitable consequence of CMC technologies (so technological determinism). Rather, the goal of an intercultural global village will require us to attend not simply to the technologies involved, but, more fundamentally, to the *social context of the use* of these technologies. In particular, a new form of *cosmopolitanism* developed among the users of these technologies—the cosmopolitanism of dual citizens in both thick local cultures and a thin global culture—would appear critical to the development of a global democracy.

This cosmopolitanism will not necessarily result simply from exposure to CMC technologies, but rather through an intentional process of education and socialization—in Aristotelian terms, through cultivating the proper human habits (*ethos* in Greek) and virtues (in the Greek sense of *arete*, our excellence as human beings) preceding the development and use of these technologies (Aristotle, *Nichomachean Ethics*, esp. Book II, 1103a14–26). Nor is the focus on virtue, habit, and excellence exclusively Western; rather, it can be found in various expressions across cultures—for example, in Buddhism and the Confucian ideal of *chün-tzu* (the authentic or profound person), as well as in recent feminist approaches.[30] Moreover, there are historical precedents for such cosmopolitanism beginning at the level of culture itself. Cultures themselves have largely worked as dynamic entities: to a greater or lesser degree and at varying speeds, most cultures of the world are in an ongoing process of losing elements of cultural habit, practice, belief, and values while simultaneously absorbing and creating new such elements, resulting in new "hybrids" that graft such elements from neighboring cultures.[31]

In the past, in parallel with such dynamic cultural dissolution/accretion, there have always been a few who have explored and adopted to "other" cultures and new cultural mixes: the cosmopolitans, citizens of the world, who have learned to live beyond the boundaries of a particular cultural domain. What is different now is not that CMC technologies are continuing this process of stirring up cultural pots, but that they are doing so on a global scale, and at a perhaps unimaginable speed (indeed, as Sandbothe (1999) makes clear, to the point of eliminating traditional notions of "time" altogether). Because of this scope and speed, it would seem that the process of cultural intermixing now requires that not just the few, but the many—anyone who desires to participate in an intercultural global society—must become cultural hybrids (synthesizing two cultures) or cultural polybrids.

In Western historical terms, an intercultural global village will require the contemporary equivalent of Renaissance women and men, where the Western Renaissance itself emerged from and expanded on precisely the extensive cultural interactions of the Medieval period (e.g., the recovery of ancient Greco-Roman science and philosophy as refined and expanded in the Muslim world, the infusion of Chinese sciences and technologies, the interactions—political, theological, and philosophical—among Muslims, Christians, and Jews, etc.). Such polybrid cosmopolitanism contrasts in particular with the uncritical "cosmopolitanism" of the cybersurfing "cultural tourist." Consistent with the values encoded by a Western culture of capitalism and commodification—values underlying and reinforced by the rapid commercialization of the Web (see especially Yoon)—the cybertourist sees "other" cultures as merely occasions for stimulation and entertainment, something like dining in an "ethnic" restaurant. One consumes something "different," the palate is mildly stimulated by difference, but then one pays the bill and goes home to the familiar. No collisions, no culture shock, no challenge to one's own most deeply-seated beliefs follow. In contrast with the polybrids and syntheses that result from leaving the cave, no enriched understanding of the whole complex of beliefs, values, views, and language(s) that make up a different culture results for the cultural tourist. Rather, "the other" is represented merely as another consumable resource, to be assimilated without resistance. As our essays show, technology and its embedded values are not the unstoppable force credited by (hard) technological determinism. But the soft determinism of a Web driven primarily by commercialization, if coupled with an uncritical ethnocentricism among those already within the cultural domain defining much of contemporary Internet and Web culture, only colludes with a cultural imperialism, the homogenization of McWorld. In contrast with the intellectual mutts and cultural polybrids necessary as dual citizens in an intercultural global village, the *Star Trek* Borg—a "culture" that consumes all the diverse cultural capital it encounters, reducing it to a single homogenous sameness—is a suggestive image of the ethnocentric cultural consumer.

As the allegory of the cave and its expression in the ancient Stoic vision of a cosmopolitan suggest, philosophy may play a crucial role in educating the dual citizens, the multicultural persons who, unlike the cultural consumers fostered by a single Internet culture, must now create senses of identity that stretch comfortably across the boundaries of multiple cultures. Philosophy works to uncover and critique the foundational assumptions defining specific cultural worldviews, and to reconstruct individual and collective worldviews that emerge from the debris of cultural collisions. More broadly, the papers collected here, along with the many other presentations and contributions to CATaC'98, trace the often obvious and sometimes subtle results of what happens when cultures collide—when Western CMC technologies are introduced into diverse cultures. A complex but coherent picture begins to emerge. The cultural collisions documented here help us uncover previously tacit assumptions about the desirability of what turns out to be, in many ways, a distinctively Western vision of the electronic global village and its collateral assumptions concerning a universal human nature, the central role of communication and communication technologies in founding and sustaining democratic polities, the neutrality (in both cultural and communicative terms) of CMC technologies, and so forth. These collisions help make explicit fundamental differences among diverse cultural values and communication styles, and they give us a much better understanding of the power and limits of contemporary CMC technologies. Perhaps most importantly, these papers also develop a trajectory towards a distinctive and hopeful model for the future of a global Internet, one which cuts between the usual dichotomies between utopia and dystopia, and between global (and potentially imperialistic) and local (and potentially isolated) cultures. In this middle ground may emerge a pluralistic humanity—dual citizens and polybrids at home in both distinctive ("thick") local cultures and a global (but "thin") on-line culture.

In David Kolb's (1998) helpful image, perhaps the fiber optics and other network technologies will become an electronic Silk Road, whose trading cities house peoples of diverse traditions and beliefs living in relative harmony with one another. The tools of CMC will allow for certain kinds of communication, but (in the short run) not all—and thus remain only one set among many of "communications suites," i.e., ways and means of communicating that individuals and groups can invoke as befits specific goals and contexts. And, contrary to the American presumptions, it seems unlikely that communication alone, no matter how facilitated by CMC technologies, will erase all conflicts between individuals and peoples. Communication here will not always be clear, either between individuals within a shared culture or cross-culturally; but, as anyone who has learned another language and lived in a different culture knows, we learn from our mistakes, especially in an environment of good will and patience with one another.

Out of this middle ground of a plurality of cultural systems and their collusions and collisions, moreover, will emerge not only "pidgins"—e.g., the sterilized airport music and the thin English often found in Internet communications—but also new and rich composites such as those noted here. But again, such composites and cultural plurality will require first of all intercultural persons, dual citizens who proceed carefully and demonstrate a deep understanding and strong respect for diverse values, traditions, customs, and beliefs. Even under these circumstances, cultural collisions are inevitable (and are in some ways the "normal" story of human history). But as initiated by such intercultural persons, we may be cautiously optimistic that the collisions and collusions mediated by global computer networks may also lead to a rich diversity of more local and more global cultures, as these cultures take up new CMC technologies.

In sum, we hope that these essays will contribute some of the insights and understandings, both theoretical[32] and practical, needed to move towards a genuinely intercultural global village—one that avoids both McWorld and Jihad as we learn to use CMC technologies in ways that globalize communication while sustaining the integrity of diverse cultural worldviews and communicative practices.

Notes

I would like to acknowledge with great gratitude several colleagues beyond the CATaC group who contributed significantly to this essay with their critical suggestions: Susan Herring (Indiana University), Caroline Reeves (Williams College), and Henry Rosemont, Jr. (St. Mary's College of Maryland). In addition to their scholarly assistance, each of these scholars exemplifies the polybrid intercultural person that I describe in my concluding remarks as necessary to a genuinely intercultural global village. I thank them for their generosity, insight, and inspiration.

I am especially delighted to acknowledge here the enormous role played by Fay Sudweeks. As co-chair of the CATaC'98 conference, she cheerfully and ably took on many of the innumerable and often daunting details of organizing a first-time international and interdisciplinary conference, with uniformly superb results. As co-editor of the several journal issues featuring papers originally presented at CATaC'98, she has been a constant source of encouragement, enthusiasm, and wise editorial judgment. Her intelligence, labor, and steady spirit have contributed to this volume in several ways, ranging from initial assistance in editorial choices to insightful suggestions and sage advice throughout the development and refinement of these essays. All who benefit from the CATaC conferences and their expressions—including this volume—owe Fay great gratitude.

Of course, I remain entirely responsible for error and poor judgments.

1. The concern that inequalities of access and power will only be amplified—rather than, as the enthusiasts promise, ameliorated—by computing technologies is not novel: see Brzezinksi (1969 [1970]). In addition to his anticipation of what is now called "the digital divide" between the haves and the have-nots, Brzezinksi also noted that as electronic communication eliminates "the two insulants of time and distance," and thereby engenders the threat that ". . . the instantaneous electronic intermeshing of mankind

will make for an intense confrontation straining social and international peace" (196).

The digital divide can be seen in several ways. As a start, consider the demographics of the Net. While estimates are admittedly inexact, in February 2000, Nua (<http://www.nua.ie/>) reported a world total of 275.54 million users. The North (Canada, Europe, and the USA) comprised 208.05 million users—some 76% of the total Internet population. Asia and the Pacific—home to more than half the world's population—totaled 54.9 million users (19.92%). South America boasted 8.79 million users (3.2%); the Middle East, 1.29 million (.47%); Africa was estimated to host 2.46 million users (.89%). This last figure is consistent with Fay Sudweeks' point: citing Tehranian (1999), she observes that there are fewer telephone lines in the entire African continent than in Tokyo (Sudweeks and Ess 1999). The World Bank's *World Development Report 1999/2000* reveals the same pattern: as of January 1999, the US claimed 1,131.52 Internet hosts per 10,000 people—compared with a world average of 75.22 and .13 for India (<http://www.worldbank.org/wdr/2000/pdfs/engtable19.pdf>: cf. Keniston in this volume).

Finally, Hoffman, Novak, and Schlosser (2000) find that the digital divide within the US is decreasing by some measures—but increasing by others (e.g., with regard to access to and use of the Web from home). For that, the global economic trend is one of increasing disparity between rich and poor. Most dramatically, "The ratio between average income of the world top 5 percent and the world bottom 5 percent increased from 78 to 1 in 1988, to 123 to 1 in 1993" (World Bank 1999). Presuming that wealth increases access to technology and infrastructure, the growing economic divide does not bode well for a putatively egalitarian global village.

2. Indeed, Carey's point can be quickly expanded by the argument that the very notion of an "electronic global village" is not simply a twentieth century, McLuhanesque dream, but rather rests precisely on Jefferson's vision of an "academical village," embodied in his designs for the University of Virginia, as a kind of education and communication system intended to expand democracy by educating a new "natural aristocracy" of young men who were to become leaders in the new republic (Wilson 1993, 71).

3. Barber writes: "The mood is that of Jihad: war not as an instrument of policy but as an emblem of identity, an expression of community, an end in itself. Even where there is no shooting war, there is fractiousness, secession, and the quest for ever smaller communities" (1992, 60). As he goes on to point out, for Muslims "Jihad" means first of all an internal spiritual struggle, and only secondarily a "holy war"; even then, Muslims insist, a Jihad should be a defensive war, not the offensive "evangelical" war connoted by Western journalistic usage of the term. Barber is careful to acknowledge that his use of the term is thus "rhetorical," one in keeping with journalistic use. But that use, I must note, is offensive to Muslims, precisely

because it reinforces a Western stereotype of Islam as a warlike religion—a stereotype that takes a very tiny number of "fundamentalists" to be representative of all Muslims. Such a stereotype is false and misleading, and runs counter to the spirit of dialogue intended here.

4. In this context, I use the term "American" as a convenient shorthand to refer specifically to the cultural mixtures and discourses characteristic of the United States.

5. *Kulturwissenschaften*—literally translated as "cultural sciences"—is distinct, however, from Cultural Studies as defined in the Anglo-American context. The German term refers specifically to the "contents and traditions of cultural analysis and cultural theory in the German-speaking world, and secondly to . . . comparable schools of thought in other academic cultures and traditions." (*IFK news* 1/99, 30).

6. The phrase "creative interferences" was introduced into our discussions at CATaC'98 by Willard McCarty. The analogy between the Silk Road and the wires and fiber holding together the Internet was part of David Kolb's closing remarks. See below, p. 28.

7. As David Kolb (1996) has already eloquently argued, we are creatures of finite time and space; as such, in the face of the exponentially increasing amount of information available on the Net, we will turn increasingly to centers and portals to help us navigate its oceans of information.

8. Indeed, several examples reported at CATaC'98 seem to exemplify this notion of a partial public: this conception seems borne out in *praxis* by the empirical examples of NGO's in Uganda (McConnell 1998) and "Celtic Men" (a men's discussion group, Rutter and Smith 1998).

9. The distinction between "high context/low content" and "high content/low context," derived from Hall (1979), is widely used in communication theory and emerges as a central theoretical element in several of our chapters. Briefly, the contrast—illustrated in this volume perhaps most dramatically by Lorna Heaton's description of CSCW systems in Japan—draws attention to communication preferences that stress the direct and efficient delivery of content, with relatively little attention to context (including the gender and relative social status of sender and recipient, their professional status vis-à-vis one another, and other socially-defined aspects of identity having to do with "face"). A standard e-mail message is a good example of high content/low context. The textual component—the majority of the information of the message—is the centerpiece; questions of gender, social status—indeed, in some cases, even the identity—of the sender and recipient are not always an obvious component of what is communicated. By contrast, "high context/low content" reverses these emphases. As Heaton and others make clear, such "high context/low content" communication is much more characteristic of Asian

and traditional societies. And in her analysis, the greatest part of the communication bandwidth of a CSCW system in Japan is devoted to conveying precisely the non-verbal modes of communication—body posture and distance, hand gesture, and gaze—that help establish the context (including relative social and professional status, etc.).

Zaharna (1995) documents the high context/low content character of communication preferences in Arabic societies. [Westerners in particular should be careful not to follow media practices of collapsing "Islam" (a religion encompassing two major traditions, Shi'i and Sunni, along with a rich and complex heritage of mysticism, poetry, architecture, etc.) and "Arabic" (a linguistic/cultural category) into other categories better attached to specific nation-states. What I refer to here as the Islamic world extends from Africa through the Middle and Far East (including Malaysia and Indonesia) to the United States, where Islam is one of the fastest growing religions today.]

10. This concept of *habitus*—at work, Lorna Heaton reminds us (in this volume), in Hofstede's analyses of cultural patterns and technology diffusion—thus seems consistent with Ihde's notion of "soft determinism" (see note 14, below).

11. To begin with, Wang (1991) found in her seven-nation survey of college students marked differences in cultural attitudes towards the ability of technology to resolve important social problems: consistent with Carey's analysis, she found that Americans were more optimistic than Asians concerning the contribution of technology to democratization. An early indication that CMC technologies themselves were not culturally neutral was provided by the reaction of several Asian countries to the possibility of introducing Internet and Web access within their boundaries. Proponents of the global village, in cosmopolitan fashion, take the values of democratic governance, individualism, and affiliated notions of human rights (including the right to free speech) as normative and legitimate for all people; indeed, a characteristic theme of especially postmodern analyses of CMC technologies is that they will inevitably extend these values as somehow intrinsic to their very design (e.g., as the distributed nature of the Internet makes censorship problematic, etc.). But these values are neither universally shared nor universally desired, as especially Asian responses to the rise of the Internet demonstrate. Singapore's reaction was characteristic: "Open the windows, but swat the flies" (Brigadier-General George Yeo, *The Straits Times*, 18 March 1995, quoted in Low 1996, 12). While the economic advantages of rapid and extensive information transfer are clearly attractive, Singapore and other Asian countries, reflecting their own deep cultural traditions, have mobilized against inadvertently importing other values seemingly characteristic of CMC technologies, including sexual permissiveness, pornography, individualism, materialism/hedonism, as well as the values of democratic polity itself.

These fears are not groundless: there is evidence that the new media do shape new, more individualized conceptions of self-identity, conceptions directly in conflict with traditional Asian worldviews (Goonasekera 1990). Singapore's effort to carefully control the information conveyed through Internet connections so as to preserve Asian cultural values against Western permissiveness, etc., is especially well documented (Low 1996; Wong 1994; Sussman 1991; on Malaysia, cf. Ang 1990). Indeed, the attitudinal differences noted by Wang (1991) are mirrored in practice; Tan et al (1998) found that CMC technologies reduced status effect in both the US and Singapore, but Singapore groups, as more conscious of status, were still able to sustain status influence. At the same time, however, a recent report on websites located in Singapore—including sites for a sex club, gay and lesbian rights, etc.—demonstrates that governmental efforts to "swat the flies" have not been entirely successful (Ho 2000). Such results, on first glance, are consistent with Deborah Wheeler's findings reported here: while CMC technologies may have a liberating effect, especially among the younger generation, the use of such technologies also mirrors prevailing cultural values. More broadly, these findings mesh with Ihde's notion of soft determinism (see note 14, below), a philosophical understanding of technologies' impact that further coheres with Keniston and Hongladarom's notions of "dual citizenship" in both "thick" local culture(s) and a global but "thin" (and thus not hegemonic) Internet culture.

12. Hongladarom (2000) helpfully summarizes Walzer's distinction by observing that "thick" morality is locally based, and is expressed in part through specific histories, narratives, and myths that help constitute a given culture's sense of identity. A "thin" culture, by contrast, can be widely shared across specific cultures because its content—including key terms such as "justice" and "truth"—is open to a wide range of interpretation and thus application in diverse contexts.

13. As Caroline Reeves (1999) points out, Robertson (1992) developed the term "glocalization" to describe the sort of synthesis and hybrid that Hongladarom develops here. See also Hongladarom (2000).

14. Several philosophers of technology, especially those concerned with the relationships between technology and democracy, have criticized technological determinism on numerous grounds. In addition to Habermas and his predecessors in the Frankfurt School (especially Marcuse 1968), the most notable include Jacques Ellul (1964), Albert Borgmann (1984), Langdon Winner (1986), Andrew Feenberg (1991), and Don Ihde (1975, 1993).

Ihde (1975) is particularly helpful here as he distinguishes between a hard and soft determinism. Using the example of the typewriter, he argues that phenomenologically, the machine cannot fully determine (hard determinism) the use of one style and the abandonment of another, ". . . but it can, through its speed, 'incline' the user away from the [*belles lettres*] style by making that style more difficult to produce" (197). On Ihde's showing, this soft determinism involves impacts of greater subtlety than impacts we would anticipate from a relationship of hard determinism between machine and user (197, 200). But we can observe here that while soft determinism thus preserves some room for individual and cultural choice in the face of new technologies, the very subtlety of technologies' impacts makes it all the more difficult to discern and anticipate these impacts, and thus to exercise choice in an informed way.

Of additional interest here, Street (1992) synthesizes many of these critiques in what he calls a "cultural approach" to technology, precisely in order to address the problem of democratic control of new technologies, control both promised and potentially frustrated by the new communications technologies (cf. Volti 1995). Both Street and Ihde (1993), moreover, counter the claims of technological determinism in part precisely by documenting how different cultures respond in different ways to technology and technological innovations. Most recently, Borgmann (1999) offers an especially powerful appreciation of the differences between natural and artificial forms of reality and information.

In the literature of communication theory, critiques of technological determinism are also developed by Ang (1990), Calabrese (1993), Venturelli (1993), Wong (1994), and Tremblay (1995). In particular, deterministic/materialist frameworks ignore the ability of individual persons to respond to and mediate larger cultural and technological influences in various ways (cf. Hall 1992, Lee et al. 1995). This ability is documented especially in Gudykunst et al. (1996), who found that individual communication preferences are not only the behavioral result of larger cultural preferences (along a spectrum of collectivist societies/low-content messages to individualist societies/high-content messages), but also correlate with individual self-construals and preferences.

As we are beginning to see (cf. notes 10, 11, above), these critiques of technological determinism in the literatures of both philosophy and communication theory are consistent with the empirical findings presented in this volume.

15. But not impossible. Beyond the efforts at such control we have already seen, it is worth noting that Saudi Arabia currently seeks to control and monitor information by relying on a single Internet node through which all communication into and out of the country must pass. This contrasts, interestingly enough, with China and other countries which, while seeking to monitor and control Internet traffic in some measure, are developing more decentralized and arguably more open infrastructures (Winship 1999).

16. See note 14 above.

17. Such middle grounds, in my view, realize the best promises of CMC technologies. As a much more modest but related example, see Ess and Cavalier (1997). Here we document our efforts at the Center for the Advancement of Applied Ethics (Carnegie Mellon University) to exploit familiar commu-

nicative advantages of CMC technologies, coupled with rules of discourse derived from Rawls and Habermas, in online dialogues that brought together participants from widely diverging points of view. Most dramatically, our online dialogue on abortion included a prominent "pro-life" Catholic spokeswoman, an active "pro-choice" Protestant minister, feminists and other ethicists. The dialogue exceeded our wildest expectations—and the common experience of face-to-face dialogues—as it indeed resulted in a remarkable consensus among the participants, including Protestant and Catholic representatives. Participants agreed, namely, that (a) abortion is not a positive good, and that our society would be improved if the demand for abortions were reduced, and (b) education could play a prominent role in helping reduce the demand for abortion. This consensus, finally, preserved irreducible differences, including those religiously-grounded differences defining Catholic and Protestant. Each acknowledged that the education programs of his or her own faith community, while aiming at the same goal of reducing abortion, would also remain distinctive as these programs would reflect, of course, the basic values and assumptions of their respective faith communities.

Michael Dahan (1998) is currently seeking to exploit CMC technologies in a similar but much more ambitious way—namely, in service to the goal of bringing together Arabs and Israelis.

18. In still other terms, as Ihde's notion of soft determinism suggests (see note 14, above), these technologies are intrinsically ambiguous with regard to their social and cultural impacts. As an additional example of such ambiguity with regard to the political dimension, Voiskounsky (1999) invokes Adorno's distinction between democratic and authoritarian personalities. He then argues that the authoritarian personality will prefer a defined path through a series of hypertext links, whereas the democratic personality will prefer maximum choice and control of what links s/he will pursue. Once again, the same technology—the hypertextual linking that is the essential structure of the web—can be taken up in two rather distinctive fashions.

This ambiguity and soft determinism thus makes possible precisely the middle grounds articulated in this volume especially by Keniston and Hongladarom between a homogenizing globalization and local identities preserved only through Jihad.

19. In addition to Ihde (1993), see Shrader-Frechette and Westra (1997) for an overview of these and other basic philosophies of technology. From a more explicitly Buddhist perspective, Herschock (1999) provides both an extensive critique of Western information technologies as not only embedding specific Western values (individuality, freedom, etc.), but also as thereby colonializing the consciousness of its users in ways that threaten both their genuine enlightenment and cultural diversity.

As well, Tagura (1997) points out that McLuhan explicitly endorsed technological instrumentalism (McLuhan 1965, 11; in Tagura 1997 ftn. 4, pp. 2f.)

In his study of ideology and technology transfer in the Philippines, Tagura argues, to the contrary, that technology—specifically communication technology—is not culturally neutral; he finds, instead, that technology is "the simultaneous bearer and destroyer of values" (Tagura 1997, 21, referring to Goulet 1977, 17–24).

Tagura's study is an excellent example of an interdisciplinary approach to questions of technology, culture, and communication. It is useful for its bibliographic resources on the core concept of "culture" as well as philosophies of technology. Tagura draws on rich philosophical (including Habermas, Winner, and Borgman) and religious (including Buddhism and Gandhi) traditions to argue for ways of encouraging economic and political development in the Philippines that flow from the values of "justice, equity, efficiency for all, cultural and ecological integrity, and the elimination of large scale systematic violence from human life." (171). Given his extensive analysis of the Philippine case, Tagura argues that this Philippine version of democratic development will require basic structural changes, beginning with changes in property ownership and relations (land reform), the development of decentralized "People's Organizations" (similar to the "base communities" of Latin America, including explicit ties to liberation theology), and greater Philippino (rather than multi-national) control over technology transfer. He acknowledges the importance of the sorts of localization of CMC technologies highlighted in this volume by Keniston: but such localization is literally a footnote in his lengthy concluding chapter (see Tagura, ftns. 49, 50, p. 205).

The point here is not simply that Tagura's study, while exemplary and useful, remains limited for our purposes insofar as his consideration of communication technology pays virtually no attention to CMC technologies and computer networks (as appears to be appropriate, given the Philippine case). At the same time, Tagura thus offers us another counterexample to the general (Western) emphases on CMC technologies as central to global democratization and economic prosperity.

20. For example, Adrie Stander (1998) helpfully documented the various cultural barriers encountered in attempting to teach computer use among students representing South Africa's many indigenous peoples, beginning with interface icons utterly meaningless outside Western cultural contexts (cf. Evers 1998). Similarly, Turk and Trees (1998) examined the conflicts between especially the epistemological assumptions built into Western information technologies and those characteristic of three indigenous peoples in Australia: see also Turk and Trees (1999).

21. Interestingly, these absences may be in part explained by some of the theoretical and practical insights garnered at CATaC'98 itself and represented in this volume. In particular, several CATaC'98 presenters referred to Hall's distinction between high content/low context (e.g., US culture and, arguably, extant CMC technologies) vis-à-vis high context/low content (e.g., Lorna Heaton's account for Japan; cf. Gill 1998). It is already

documented that Arabic cultures are usefully characterized as high context/low content; hence, while the Internet and the Web are clearly spreading rapidly throughout the Islamic world—witness Deborah Wheeler's contribution to this volume—there may be a greater cultural mismatch between communication preferences and values in Arabic-speaking cultures and CMC technologies than found among the communication preferences and values of cultures more fully represented at CATaC'98 (see Hall 1979; Zaharna 1995).

22. In doing so, Lacroix and Tremblay observe that the non-material character of norms and values frees them from the reductive materialism and determinism presumed in many theoretical approaches: in particular, values and norms are collectively "owned" and, in that sense, freely available. This frees cultural norms and values from both the commodification and utilitarian calculations based on scarcity which they see centrally at work in the "culture industries" (1997, 41).

More broadly, the non-material character of norms and values opens up precisely the possibility of choice, both individual and collective, in the face of the otherwise overwhelming power of culture and technology. Such choice, as we have seen, is apparent in multiple ways and is theoretically included in the notion of *habitus* as elaborated here especially by Yoon.

23. In addition to Lacroix and Tremblay's recognition of the difficulties of defining "culture," see Star (1995, 26). While her anthology, *The Cultures of Computing,* includes a number of significant and pertinent essays, her focus is on culture in a related but different sense than we use it here. That is, Star seeks to "talk about a set of practices with symbolic and communal meaning" (26)—i.e., as in the example of "organizational culture," the "transnational culture" of professionals from different countries working together via e-mail, etc. For our part, "symbolic and communal meaning" is certainly of central interest, but primarily as affiliated with national and linguistic boundaries, communication preferences, etc.

Moreover, in light of Moon's meta-analysis of how the term "culture" is used in intercultural communication, it would appear that our difficulty in identifying a clear understanding of "culture" reflects changing understandings and debate within the field of intercultural communication itself (1996). She notes:

> As a rule, intercultural communication scholars are not interested in the idea of "culture" per se, but use operationalized notions of cultural variation (e.g., individualism/collectivism) as one among many independent variables that affect the dependent variable. . . . "Culture," at this level, is most often defined as nationality, and the constructedness of this position and its intersection with other positions such as gender and social class is not considered. The outcome is that diverse groups are treated as homogenous, differences within national boundaries, ethnic groups, genders, and races are

> obscured, and hegemonic notions of "culture" are presented as "shared" by all cultural members. (76)

To correct these deficits, Moon argues for taking up critical and feminist perspectives which would "allow intercultural communication scholars to employ more sophisticated and politicized analyses of cultural identity in general and to examine how these identities are constructed in communication, as well as how they affect communication" (76). Similarly, Martin and Flores (1998), in their overview of contemporary paradigms in communication theory concerning culture and communication, call for an "interparadigmatic dialogue" to further this study—one that, echoing Moon, calls for the insights of postmodern and feminist scholarship, as well as critical theory. The essays collected here are partial responses to these calls for additional understandings of culture, not only within the postmodernist and feminist frames, but other frames as well (including Habermas, hermeneutics, and others).

Pasquali (1985) helpfully explores the possibility of a "philosophy of culture" appropriate to especially mass communication technologies (i.e., as distinct in important ways from CMC technologies). For other discussions of culture, including its relation to language, see Singer (1987), Rosengrun (1994), and Garcea (1998).

24. For discussion of these and related issues in the debates between Habermas and Foucault, see Kelly (1994), d'Entrèves et al. (1997), and Sawicki (1994).

25. For my own sketches of the complementarities between philosophy and communication theory, see Ess (1996), and Ess (1999).

26. Consider, for example, Barlow's definition of cyberspace: "Cyberspace consists of transactions, relationships, and thought itself, arrayed like a standing wave in the web of our communications. Ours is a world that is both everywhere and nowhere, but it is not where bodies live." This implicitly Gnostic dualism, and its hostility towards the material world (which Barlow refers to contemptuously as "meatspace," in contrast with cyberspace) leads to a complete rejection of the material order, including the legal system:

> Your legal concepts of property, expression, identity, movement, and context do not apply to us. They are based on matter, There is no matter here. Our identities have no bodies, so, unlike you, we cannot obtain order by physical coercion. . . . We must declare our virtual selves immune to your sovereignty, even as we continue to consent to your rule over our bodies. We will spread ourselves across the Planet so that no one can arrest our thoughts. We will create a civilization of the Mind in Cyberspace. (1996)

In contrast with the cybergnostic enthusiasm for escaping body in virtual community, others have also documented the role of embodiment in anchoring us in a real world of diverse cultures and communities: see Argyle and Shields (1996); Baym (1995); and Bromberg (1996).

27. As Vivian Sobchack observes, Haraway moves away from her earlier optimism that women's liberation would be best accomplished through abandoning the body: Sobchack makes this point in her powerful critique of the contempt for the body as "meat" characteristic of Barlow, the postmodernist Baudrillard and others (1995).

28. The term "culture shock" derives from Oberg's seminal article (1960), included in a useful collection of chapters (some classic) on "Culture and Communication" in Weaver (1998a). See as well Weaver's own discussion (1998b).

Bennett (1977) discusses culture shock as one form of what she describes as a more general "transition shock." She does so primarily from the perspective of communication theory and psychology: her account nicely complements the very brief one I've given here in terms of worldview, a notion shared among philosophy and the social sciences. Gudykunst and Kim (1984) begin with Bennett's account as they develop their own suggestions for "Becoming Intercultural" (ch. 14).

It is by no means clear, however, that "intercultural communication training," in its current state, is fully prepared to offer us either theories or practices that always succeed in helping us become intercultural: for a review of literature and critique of prevailing models, along with their own suggestion for a new model of intercultural communication training, see Cargile and Giles (1996).

In any case, Chen and Starosta (1996) also offer what they claim to be a synthesis of earlier models, one involving three elements, including a cognitive "Intercultural Awareness" (364ff.). This element explicitly intersects with a shared focus on worldview: as philosophy and the social sciences make the various elements of worldview more explicit, they can directly contribute to such intercultural awareness.

Finally, Yuan (1997) provides a model for intercultural communication that is especially striking for its effort to synthesize an explicitly philosophical theory (Donald Davidson's philosophy of Externalism) with a rhetorical theory (Thomas Kent's theory of Paralogic hermeneutics) and the Eastern philosophical/religious perspective of Taoism, with its stress on complementarity.

29. In the Western context, consider such stories as *The Epic of Gilgamesh*, Homer's *Odyssey*, and the so-called "Adam and Eve" story (the second Genesis creation story) in Genesis 2.4b–4.26 (see Ess 1995). For an Eastern example, consider the account of the Buddha's enlightenment (e.g., Strong 1995), and compare this with Plato's account of enlightenment in the allegory of the cave (*Republic*, Book VII, 514a–517b, in Bloom 1991, 193–97) and his *Seventh Letter* (340–41), in Hamilton 1978, 134–36).

30. Boss provides a helpful introduction and selection of readings on virtue ethics and their applications, making these similarities clear as well

as contrasting them with other ethical approaches, likewise accounted for from a multicultural perspective (1999, esp. 36–41). For a more extensive introduction to ethics from cross-cultural perspectives, see Gupta and Mohanty (2000), Part 3, and Hooke (1999). May, Collins-Chobanian, and Wong (1998) is a useful anthology of ethical resources from diverse cultural sources, as applied to specific ethical issues.

31. The point that "culture" does not refer to a fixed, monolithic entity is made humorously by Ralph Linton (1937), as he describes a "One Hundred Percent American" whose daily life in fact involves a rich collage of cultural inventions and developments, e.g., from East India (pajamas), Asia Minor (bed, wool, milk), India (cotton, steel, umbrella), China (silk, porcelain, printing, paper), ancient Egypt (glass, shaving), the Near East (glazed tile, chair), Turkey (towel, coffee) ancient Rome (bathtub, toilet), the ancient Gauls (soap), etc. Linton notes that the "authentically American costume of gee string and moccasins," while more comfortable, is not likely to be a choice of attire. And, "As he scans the latest editorial pointing out the dire results to our institutions of accepting foreign ideas, he will not fail to thank a Hebrew God in an Indo-European language that he is a one hundred percent (decimal system invented by the Greeks) American (from Americus Vespucci, Italian geographer)."

Similarly, Joseph Needham's monumental investigation into the development of Chinese science and technology demonstrates that what Westerners think of as a distinctively Western natural science is in fact rooted in significant ways in China (1954–). Finally, in their anthology of primary sources, Tweed and Prothero (1999) provide a detailed history of the cultural interactions between Asian religious traditions and American culture from 1784 to contemporary interreligious dialogue and legal disputes. Along these lines, see also Eck (1997).

32. One of the central themes of subsequent CATaC conferences will be just that of developing meta-theoretical overviews and syntheses which further the sorts of interdisciplinary projects and dialogues represented here. Perhaps a sort of "super-science" ("Super-Wissenschaft"), for example, as envisioned by Wunberg (1999), will emerge. At CATaC '98 our first conversations about such meta-theories recognized a range of possibilities—from a (modernist) model of a single meta-theory which organized diverse disciplines from a single center to a (postmodernist) model of theoretical fragments held together only loosely in constantly changing, decentered, and ad hoc fashion.

Indeed, as Halloran (1986, 55), reminds us, the natural sciences do not (as yet) enjoy such theoretical unity: it is perhaps unrealistic to hope for—much less, hold dogmatically to—a single theory. In a middle ground of theoretical pluralism, multiple models and theories may be taken up—including a model suggested by biological colonies (say, those of corals) which consist of stable structures open to new growth and development (cf. Porra 1999).

References

Adler, Peter S. 1977. "Beyond Cultural Identity: Reflections on Multiculturalism." In *Culture Learning: Concepts, Applications, and Research*, ed. Richard W. Brislin, 24–41. Honolulu: University of Hawaii Press. Reprinted in Bennett (1998), 225–245.

Ang, Ien. 1990. "Culture and Communication: Towards an Ethnographic Critique of Media Consumption in the Transnational Media System." *European Journal of Communication* 5 (2–3, June): 239–260.

Argyle, Katie, and Rob Shields. 1996. "Is there a Body in the Net?" In *Cultures of Internet: Virtual Spaces, Real Histories, Living Bodies*, ed. Rob Shields, 58–69. Thousand Oaks, CA: Sage.

Aristotle. 1968. *The Nichomachean Ethics*. Trans. H. Rackham. Cambridge, MA: Harvard University Press.

Balsamo, Anne. 1998. "Myths of Information: The Cultural Impact of New Information Technologies." In *The Information Revolution: Current and Future Consequences,* eds. Alan L. Porter and William H. Read, 225–235. Greenwich, CT: Ablex.

Barber, Benjamin. 1992. "Jihad vs. McWorld." *The Atlantic Monthly*, March, 53–63.

———. 1995. *Jihad versus McWorld.* New York: Times Books.

Barlow, John Perry. 1996. "A Declaration of the Independence of Cyberspace." <http://www.eff.org/pub/Censorship/Internet_censorship_bills/barlow_0296.declaration>

Baym, Nancy K. 1995. "The Emergence of Community in Computer-Mediated Communication." In *CyberSociety: Computer-Mediated Communication and Community*, ed. Steven G. Jones, 138–63. Thousand Oaks, CA: Sage.

Bennett, Janet M. 1977. "Transition Shock: Putting Culture Shock into Perspective." *International and Intercultural Communication Annual*, vol. 4. Falls Church, VA: Speech Communication Association, December. Reprinted in Bennett (1998), 215–23.

Bennett, Milton J. 1998. *Basic Concepts of Intercultural Communication: Selected Readings.* Yarmouth, ME: Intercultural Press.

Bolter, J. David. 1986. *Turing's Man: Western Culture in the Computer Age*. Chapel Hill: University of North Carolina Press.

———. 1991. *Writing Space: The Computer, Hypertext, and the History of Writing.* Hillsdale, NJ: Erlbaum.

Borgmann, Albert. 1984. *Technology and the Character of Contemporary Life*. Chicago: University of Chicago Press.

———. 1999. *Holding onto Reality: The Nature of Information at the Turn of the Millennium*. Chicago: University of Chicago Press.

Boss, Judith. 1999. *Analyzing Moral Issues*. Mountain View, CA: Mayfield.

Brzezinski, Zbigniew K. 1969 [1970]. "The Search for Meaning Amid Change." *New York Times*, C141+, January 6; reprinted in *Computers and Society*, ed. George A. Nikolaieff, 190–96. New York: H.W. Wilson, 1970.

Bromberg, Heather. 1996. "Are MUDs Communities? Identity, Belonging and Consciousness in Virtual Worlds." In *Cultures of Internet: Virtual Spaces, Real Histories, Living Bodies*, ed. Rob Shields, 143–52. Thousand Oaks, CA: Sage.

Calabrese, Andrew. 1993. "Designing Communication: The Culture and Politics of the Electronic Cottage." *Progress in Communication Sciences* 11: 75–100.

Carey, James. 1989. *Communication as Culture: Essays on Media and Society*. Boston: Unwin Hyman.

Cargile, Aaron Castelan, and Howard Giles. 1996. "Intercultural Communication Training: Review, Critique, and a New Theoretical Framework." *Communication Yearbook* 19: 385–423.

Chen, Guo-Ming, and William J. Starosta. 1996. "Intercultural Communication Competence: A Synthesis." *Communication Yearbook* 19: 353–83.

Crowley, David, and Paul Heyer, eds. 1996. *Communication in History: Technology, Culture, Society*. 2nd ed. White Plains, N.Y.: Longman.

Chesebro, James W., and Dale A. Bertelsen. 1996. *Analyzing Media: Communication Technologies as Symbolic and Cognitive Systems*. New York: Guilford Press.

Dahan, Michael. 1998. "National Security and Democracy on the Internet in Israel." In *Proceedings: Cultural Attitudes Towards Technology and Communication,* eds. Charles Ess and Fay Sudweeks, 145–58. Sydney: Key Centre of Design Computing.

Eck, Diana, and the Pluralism Project at Harvard University. 1997. *On Common Ground: World Religions in America*. (CD-ROM). New York: Columbia University Press.

Ellul, Jacques. 1964. *The Technological Society*. New York: Alfred A. Knopf.

d'Entrèves, Maurizio Passerin, and Seyla Benhabib. 1997. *Habermas and the Unfinished Project of Modernity: Critical Essays on the Philosophical Discourse of Modernity*. Cambridge, MA: MIT Press.

The Epic of Gilgamesh. 1985. Trans. Maureen Gallery Kovacs. Stanford, CA: Stanford University Press.

Ess, Charles. 1995. "Reading Adam and Eve: Re-Visions of the Myth of Woman's Subordination to Man." In *Violence Against Women and Children: A Christian Theological Sourcebook*, eds. Carol J. Adams and Marie M. Fortune, 92–120. New York: Continuum.

———. 1996. "Thoughts Along the I-Way: Philosophy and the Emergence of Computer-Mediated Communication." In *Philosophical Perspectives on Computer-Mediated Communication*, ed. Charles Ess, 1–12. Albany: State University of New York Press.

———. 1998. "First Looks: CATaC'98." In *Proceedings: Cultural Attitudes Towards Technology and Communication,* eds. Charles Ess and Fay Sudweeks, 1–17. Sydney: Key Centre of Design Computing. Available on-line: <http://www.arch.usyd.edu.au/~fay/catac/index.html>.

———. 1999. "Critique in Communication and Philosophy: An Emerging Dialogue." Review of James W. Chesebro and Dale A. Bertelsen, *Analyzing Media: Communication Technologies as Symbolic and Cognitive Systems* (New York: Guilford Press, 1996). *Research in Philosophy and Technology* 18: 219–26.

Ess, Charles, and Robert Cavalier. 1997. "Is There Hope for Democracy in Cyberspace?" In *Technology and Democracy: User Involvement in Information Technology*, eds. David Hakken and Knut Haukelid, 93–111. Oslo: Center for Technology and Culture.

Evers, Vanessa. 1998. "Cross-Cultural Understanding in Interface Design." In *Proceedings: Cultural Attitudes Towards Technology and Communication*, eds. Charles Ess and Fay Sudweeks, 217–18. Sydney: Key Centre of Design Computing.

Feenburg, Andrew. 1991. *Critical Theory of Technology*. Oxford: Oxford University Press.

Finn-Jordan, Kathleen A. 1998. "Third Culture Persons." In *Culture, Communication and Conflict*, ed. Gary R. Weaver, 242–49. New York: Simon and Schuster.

Garcea, Elena A. A. 1998. "European Perspectives on Intercultural Communication." *European Journal of Intercultural Studies* 9 (1): 25–40.

Gates, Bill, with Nathan Myhrvold and Peter Rinearson. 1996. *The Road Ahead*. New York: Penguin Books.

Gill, Satinder. 1998. "The Cultural Interface: The Role of Self." In *Proceedings: Cultural Attitudes Towards Technology and Communication*, eds. Charles Ess and Fay Sudweeks, 202–7. Sydney: Key Centre of Design Computing.

Gudykunst, William B., and Young Yun Kim. 1997. *Communicating with Strangers: An Approach to Intercultural Communication*. 3rd ed. New York: McGraw-Hill.

Gudykunst, William B., Yuko Matsumoto, Stella Ting-Toomey, Tsukasa Nishida, Kwangsu Kim, and Sam Heyman. 1996. "The Influence of Cultural Individualism-Collectivism, Self-Construals, and Individual Values on Communication Styles Across Cultures." *Human Communication Research* 22, no. 4 (June): 510–43.

Gupta, Bina, and J. N. Mohanty. 2000. *Philosophical Questions: East and West*. New York: Rowman and Littlefield.

Hall, Bradford 'J'. 1992. "Theories of Culture and Communication." *Communication Theory* 2, no.1 (February): 50–70.

Hall, Edward T. 1979. "Learning the Arabs' Silent Language" (interview by Kenneth Friedman), in Weaver (1998a), pp. 17–22. Reprinted from *Psychology Today*, 13, no. 3 (August).

Halloran, James D. 1986. "The Social Implications of Technological Innovations in Communication." In *The Myth of the Information Revolution: Social and Ethical Implications of Communication Technology,* ed. M. Traber, 7–20. London: Sage.

Hamelink, C. J. 1986. "Is There Life After the Information Revolution?" In *The Myth of the Information Revolution: Social and Ethical Implications of Communication Technology*, ed. M. Traber, 46–63. London: Sage.

Haraway, Donna J. 1990. "A Cyborg Manifesto: Science, Technology, and Socialist-Feminism in the Late Twentieth Century." In *Simians, Cyborgs, and Women: The Reinvention of Nature*, 149–81. New York: Routledge.

Herring, Susan C. 1996. "Posting in a Different Voice: Gender and Ethics in Computer-Mediated Communication." In *Philosophical Perspectives on Computer-Mediated Communication*, ed. Charles Ess, 115–45. Albany: State University of New York Press.

———. 1999a. "The Rhetorical Dynamics of Gender Harassment On-line." *The Information Society* 15 (3): 151–67.

———. 1999b. Personal communication.

Herschock, Peter D. 1999. *Reinventing the Wheel: A Buddhist Response to the Information Age*. Albany: State University of New York Press.

Ho, Kong Chong. 2000. "Sites of Resistance: Charting the Alternative and Marginal Websites in Singapore." Internet Research 1.0: The State of the Interdiscipline (First Conference of the Association of Internet Researchers), Lawrence, Kansas, USA, September 14.

Hoffman, Donna L., Thomas P. Novak, and Ann E. Schlosser. 2000. "The Evolution of the Digital Divide: How Gaps in Internet Access May Impact Electronic Commerce." *Journal of Computer-Mediated Communication* 5 (3). Available on-line <http://www.ascusc.org/jcmc/vol5/issue3/hoffman.html>

Homer. 1963. *The Odyssey.* Trans. Robert Fitzgerald. New York: Anchor Books.

Hongladarom, Soraj. 2000. "Negotiating the Global and the Local: Some Philosophical Reflections on the Impact of the Internet on Local Cultures." Eighth East-West Philosophers' Conference, University of Hawai'i at Manoa, Honolulu, Hawaii, January 12.

Hooke, Alexander E. 1999. *Virtuous Persons, Vicious Deeds.* Mountain View, CA: Mayfield.

Ihde, Don. 1975. "A Phenomenology of Man-Machine Relations." In *Work, Technology, and Education: Dissenting Essays in the Intellectual Foundations of American Education,* eds. W. Feinberg and H. Rosemont, Jr., 186–203. Chicago: University of Illinois Press.

———. 1993. *Philosophy of Technology: An Introduction.* New York: Paragon House.

Jones, Steve. 1998. Closing comments, CATaC'98 (Science Museum, London, UK), August 3.

Kelly, Michael. 1994. *Critique and Power: Recasting the Foucault/Habermas Debate*. Cambridge, MA: MIT Press.

Kolb, David. 1996. "Discourse Across Links." In *Philosophical Perspectives on Computer-Mediated Communication*, ed. Charles Ess, 116–26. Albany: State University of New York Press.

Lacroix, Jean-Guy, and Gaëtan Tremblay. 1997. "Trend Report: The 'Information Society' and Cultural Industries Theory." *Current Sociology* 45, no.4 (October): 1–162.

Landow, George. 1992. *Hypertext: The Convergence of Contemporary Critical Theory and Technology*. Baltimore: Johns Hopkins University Press.

Landow, George, ed. 1994. *Hyper/Text/Theory*. Baltimore: Johns Hopkins University Press.

Lee, Wen Shu, Jianglong Wang, Jensen Chung, and Ellen Hertel. 1995. "A Sociohistorical Approach to Intercultural Communication." *Howard Journal of Communications* 6, no. 4 (December): 262–91.

Lievrouw, Leah A. 1998. "Our Own Devices: Heterotopic Communication, Discourse, and Culture in the Information Society." *The Information Society* 14: 83–96.

Linton, Ralph. 1937. "One Hundred Per Cent American." *The American Mercury* 40: 427–29.

Low, Linda. 1996. "Social and Economic Issues in an Information Society: A Southeast Asian Perspective." *Asian Journal of Communication* 6 (1): 1–17.

Lyotard, Jean-François. 1984. *The Postmodern Condition: A Report on Knowledge*. Trans. Geoff Bennington and Brian Massumi. Minneapolis: University of Minnesota Press.

Marcuse, Herbert. 1968. *One-Dimensional Man*. Boston: Beacon Books.

Martin, Judith N., and Lisa A. Flores. 1998. "Colloquy: Challenges in Contemporary Culture and Communication Research." *Human Communication Research* 25, no. 2 (December): 293–99.

May, Larry, Shari Collins-Chobanian, and Kai Wong. 1998. *Applied Ethics: A Multicultural Approach*. 2nd ed. Upper Saddle River, New Jersey: Prentice Hall.

McCarty, Willard. 1998. Closing comments. CATaC'98 (Science Museum, London, UK), August 3.

Mitchell, William J. 1995. *City of Bits: Space, Place, and the Infobahn*. Cambridge, MA: MIT Press.

Moon, Dreama G. 1996. "Concepts of 'Culture': Implications for Intercultural Communication Research." *Communication Quarterly* 44, no. 1 (Winter): 70–84.

Needham, Joseph. 1954–. *Science and Civilisation in China*. Cambridge: Cambridge University Press.

Negroponte, Nicholas. 1995. *Being Digital*. New York: Knopf.

Oberg, Kalvero. 1960. "Culture Shock and the Problem of Adjustment in New Cultural Environments." *Practical Anthropology* vol. 1. Reprinted in Weaver (1998a), 185–86.

Pasquali, Antonio. 1985. "Is a Philosophy of Culture Still Possible?" In *Culture in the Electronic Age*. [SERIES: Cultures, dialogue between the peoples of the world], 15–22. Paris: [New York: UNESCO; Unipub, US distributor].

Pargman, Daniel. 1999. "Reflections on Cultural Bias and Adaptation." *Javnost/The Public* VI (4): 23–37.

Plato. 1978. *Phaedrus and Letters VII and VIII*. Trans. Walter Hamilton. New York: Penguin.

———. 1991. *The Republic of Plato*, 2nd ed. Trans. Allan Bloom. New York: Basic Books.

Porra, Jaana. 1999. "Colonial Systems." *Information Systems Research* 10, no. 1 (May): 38–69.

Reeves, Caroline. 1999. Personal communication.

Richards, Cameron. 1998. "CMCs and the Problem of 'Grounding' Virtual Utopias." In *Proceedings: Cultural Attitudes Towards Technology and Communication,* eds. Charles Ess and Fay Sudweeks, 129–40. Sydney: Key Centre of Design Computing.

Robertson, Roland. 1992. *Globalization: Social Theory and Global Culture.* London: Sage.

Rosengrun, Karl Erik. 1994. "Culture, Media and Society: Agency and Structure, Continuity and Change." In *Media Effects and Beyond: Culture, Socialization and Lifestyles,* ed. K. Rosengrun, 3–28. New York: Routledge.

Samovar, Larry A. and Richard E. Porter. 1988. *Intercultural Communication: A Reader*. 6th ed. Belmont, CA: Wadsworth.

Sandbothe, Mike. 1999. "Media Temporalities of the Internet: Philosophies of Time and Media in Derrida and Rorty." *AI and Society* 13: 421–34.

Sawicki, Jana. 1994. "Foucault and Feminism: A Critical Reappraisal." In *Critique and Power: Recasting the Foucault / Habermas Debate*, ed. Michael Kelly, 347–64. Cambridge, MA: MIT Press.

Shrader-Frechette, Kristen, and Laura Westra, eds. 1997. *Technology and Values*. New York: Rowman and Littlefield.

Singer, Marshall R. 1987. "The Role of Culture and Perception in Communication." In *Intercultural Communication: A Perceptual Approach*, ed. Marshall Singer, 28–53. 2nd ed. New York: Prentice-Hall.

Smith, Peter B., Shaun Dugan, and Fons Trompenaars. 1996. "National Culture and the Values of Organizational Employees: A Dimensional Analysis Across 43 Nations." *Journal of Cross-Cultural Psychology* 27 (2): 231–64.

———. 1997. "Locus of Control and Affectivity by Gender and Occupational Status: A 14 Nation Study." *Sex Roles* 36: 1/2, 51–77.

Sobchack, Vivian. 1995. "Beating the Meat/Surviving the Text, or How to Get Out of This Century Alive." In *Cyberspace / Cyberbodies /*

Cyberpunk: Cultures of Technological Embodiment, eds. Mike Featherstone and Roger Burrows, 205–14. London: Sage Publications.

Stander, Adrie. 1998. "Bridging the Gap: Issues in the Design of Computer User Interfaces for Multicultural Communities." In *Proceedings: Cultural Attitudes Towards Technology and Communication*, eds. Charles Ess and Fay Sudweeks, 211–16. Sydney: Key Centre of Design Computing.

Star, Susan Leigh, ed. 1995. *The Cultures of Computing*. Cambridge, MA: Blackwell.

Stoll, Clifford. 1995. *Silicon Snake Oil: Second Thoughts on the Information Highway*. London: Pan Books.

Stone, Gerald, Michael Singletary, and Virginia P. Richmond. 1999. *Clarifying Communication Theories: A Hands-On Approach*. Ames: Iowa State University Press.

Street, John. 1992. *Politics and Technology*. New York: Guilford Press.

Strong, John S. 1995. *The Experience of Buddhism: Sources and Interpretations*. Belmont, CA: Wadsworth.

Sudweeks, Fay and Charles Ess. 1999. "Global Cultures: Communities, Communication, and Transformation." *Javnost/The Public* VI (4): 5–10.

Sussman, Gerald. 1991. "The 'Tiger' from Lion City: Singapore's Niche in the New International Division of Communication and Information." In *Transnational Communications: Wiring the Third World*, eds. G. Sussman and John Lent, 279–308. Newbury Park, CA: Sage Publications.

Tagura, Pablito M. 1997. *A Critical Analysis of Technology Transfer, Ideology, and Development in the Context of Communication Technology in the Philippines*. Ph.D. diss., Marquette University, Milwaukee, Wisconsin.

Tan, Bernard C. Y., Kwok-Kee Wei, Richard T. Watson, and Rita M. Walczuch. 1998. "Reducing Status Effects with Computer-Mediated Communication: Evidence from Two Distinct National Cultures." *Journal of Management Information Systems* 15, no. 1 (summer): 119–41.

Tehranian, Majid. 1999. *Global Communication and World Politics: Domination, Development and Discourse*. Boulder, CO: Lynne Reiner Publishers.

Tremblay, Gaëtan. 1995. "The Information Society: From Fordism to Gatesism." *Canadian Journal of Communication* 20, no. 4 (autumn): 461–82.

Turk, Andrew, and Kathryn Trees. 1998. "Culture and Participation in the Development of CMC: Indigenous Cultural Information System Case Study." In *Proceedings: Cultural Attitudes Towards Technology and*

Communication, eds. Charles Ess and Fay Sudweeks, 219–23. Sydney: Key Centre of Design Computing.

———. 1999. "Appropriate Computer-Mediated Communication: An Australian Indigenous Information System Case Study. *AI and Society* 13: 377–88.

Tweed, Thomas A., and Stephen Prothero. 1999. *Asian Religions in America: A Documentary History*. New York: Oxford University Press.

Venturelli, Shalini S. 1993. "The Imagined Transnational Public Sphere in the European Community's Broadcast Philosophy: Implications for Democracy. *European Journal of Communication* 8, no. 4 (December): 491–518.

Volti, Rudi. 1995. *Society and Technological Change*. 3rd ed. New York: St. Martin's Press.

Voiskounsky, Alexander. 1999. "Internet: Culture, Diversity and Unification. *Javnost: The Public* VI (4): 53–65.

Wang, Georgette. 1991. "Information Society in Their Mind: A Survey of College Students in Seven Nations." *Asian Journal of Communication* 1 (2): 1–18.

Weaver, Gary R., ed. 1998a. *Culture, Communication and Conflict*. New York: Simon and Schuster.

———. 1998b. "Understanding and Coping with Cross-Cultural Adjustment Stress." In Weaver (1998a), 187–204.

Wilson, Richard Guy. 1993. "Jefferson's Lawn: Perceptions, Interpretations, Meanings." In *Thomas Jefferson's Academical Village: The Creation of an Architectural Masterpiece*, ed. Richard Guy Wilson, 47–71. Charlottesville: University of Virginia Press.

Winner, Langdon. 1986. *The Whale and the Reactor*. Chicago: University of Chicago Press.

Winship, James. 1999. "The Internet and the Web in the People's Republic of China." ASIANetwork Conference, Tacoma, Washington, April 24.

Wong, Kokkeong. 1994. "Media, Culture, and Controlled Commodification: The Case of Peripheral Singapore." *Advances in Telematics* 2: 144–65.

Wunberg, Gotthart. 1999. Kulturwissenschaften—Geschichte und Selbstverständnis ["Cultural Analysis and Theory: History and Self-Understanding"], *news: Mitteilungen des Internationalen Forschungszentrums Kulturwissenschaften*, ed. Ines Steiner, 4–5. Vienna: Verein Internationalen Forschungszentrums Kulturwissenschaften. 1/99.

Yuan, Rue. 1997. "Yin/Yang Principle and the Relevance of Externalism and Paralogic Rhetoric to Intercultural Communication." *Journal of Business and Technical Communication* 11, no. 3 (July): 297–320.

Yoon, Sunny. 1996. "Power Online: A Poststructuralist Perspective on CMC." In *Philosophical Perspectives on Computer-Mediated Communication*, ed. Charles Ess, 171–96. Albany: State University of New York Press.

World Bank. 1999. "Income Poverty: Trends in Inequality." <http://www.worldbank.org/poverty/data/trends/inequal.htm>

Zaharna, R. S. 1995. "Understanding Cultural Preferences of Arab Communication Patterns." *Public Relations Review* 21, no. 3 (fall): 241–55.

I. Theoretical Approaches: Postmodernism, Habermas, Luhmann, Hofstede

Understanding Micropolis and Compunity

Steve Jones

In my book *CyberSociety: Computer-Mediated Communication and Community* (Jones 1995), I argued that terms commonly used in the US to describe the Internet such as "information highway" and "national information infrastructure" are unfortunate but telling metaphors. They bring with them much intellectual and social baggage, largely due to the startling parallels between the current project, this "information superhighway," and the one spurred on in the US by both World Wars, the interstate highway system—not the least of which is the reliance on the word "highway" and the romantic connotations of the open road. Another important parallel is the military origin of highway building [as established by Thomas Jefferson, among others (Patton 1986)] and the military origins of what is presently the most prominent information highway, the Internet, in Defense Department computer networks linked to university research centers. And yet another parallel is to the 1960s "space race" and our quest to lead in new technologies and science.

And race ahead we do. I think racing, to push the motoring metaphor, serves well to characterize a social bias based, in essence, on movement itself. We can acknowledge several things that compose it; competitive spirit perhaps, a modern need for mobility also, and curiosity as well. It is a movement based on speed, rooted in transportation, and oblivious in large part to that which is transported. To put it another way, loyalty is to the movement of something (often ourselves, but not always) from one place to another, to flow, and not to that which is being moved (the last word's *double-entendre* intended), to content.

I believe this quest for movement is well-illustrated by our early understanding of electricity, and can be most easily recognized in the work of Nikola Tesla (Cheney 1981). In the late 1890s Tesla envisioned a world linked by electricity. He proposed the development of a global electrical network to facilitate communication. Tesla believed that anything could be coded into electrical impulses and transmitted via electricity. In that sense he presaged the current trend toward digitization. But one might say that he also foresaw the postmodern shift from meaning to Deleuze and Guattari's (1980) concept regarding flow, from a social space within which signs took shape, metamorphosed, disappeared and reappeared, to a space where meaning shifts while signs remain. Meaning itself is fluid, mobile, and nothing should have meaning for long.

Another reason I find our use of the highway metaphor unfortunate is that it leaves aside the issue of power: it focuses our attention on the road, the infrastructure, and away from the people and "vehicles" that traverse it, away from the road-side, away from the interaction of road and place. It focuses our attention away from the gaze of others, the sense that we are as surveilled as we are social (Foucault 1977). We are led to believe we are in power, we are the ones "surfing," or "using," and others cannot see us, just as we cannot be seen when we watch television. The seeming absence of the other focuses away from economic and political issues, and directs us toward ourselves.

But there is evidence of the "other" on-line. Perhaps a metaphor from boating would serve better than on based on automobile transport. As we travel along an information "path," we leave behind a wake, though we may not leave behind tangible and permanent markers. One of the earliest discoveries in electromagnetics was that as an electrical current flows in a wire a magnetic field is generated around that wire at a right angle. The forces not only interact, they are dependent on each other, and the wire's "content," the movement of electrons through it, creates a "field" of force around it. The creation of those fields is itself dependent on movement. Such may be the case with messages we send via Internet (or for that matter via other media as well); they travel from place to place but also create a "field" of influence and meaning around themselves.

Many others (McLuhan 1965; Carey 1989; Ong 1982; Eisenstein 1979; Goody 1986) have assayed this territory, but perhaps it is necessary to do so again, as we have become far more savvy media users and producers. McLuhan's once oft-repeated phrase "the medium is the message" contains a new twist. We are not interested in the message *per se*; we are interested in getting the message across. We have less interest in what we mean and more interest in how we mediate what we say. What medium shall I use, and what will the consequences be of my choice?

Carey (1989) links the study of communication to the study of social relations, noting two trajectories along which we think about communication. The first trajectory is along the lines of the "transportation" metaphor of communication. In this model communication is, in the main, the movement of messages from one place to another. This is the model I have thus far characterized, and the model on which the communication industry itself is built.

Carey contrasts the transportation model to the "ritual" model of communication, the latter intended to connote communication as the sharing of ideas and beliefs. Whether for a particular purpose or not, whether for transmission of information or participation in those activities that make us human, be they mundane or special, the ritual model points out that communication is the medium within which we exist, as much as is the air we breathe. Again we find a twist on McLuhan—the medium is the message because the medium is not one of communication *per se* but rather it is the ground in which human connectedness can grow and flourish.

But the ritual model does not enter into our public conversation about new media, and it does not fit industry models and methods of communication technology development. To put it another way, when one is asked "Did you hear?" these days, the question connotes something about whether we are connected, wired. Forster's admonition that we "only connect" has been taken too literally. Rarely does being connected anymore carry the connotations of community, gossip, storytelling. What is connoted is instead "compunity," a merger of computers with communities and our sense of community. We long for the community and communion that the ritual model holds dear as these are elements inseparable from communication, but we are given instead the ability to send messages to and fro as disconnected and disembodied texts. The ritual model emphasizes that communication is the means by which we build our understanding of the world and ourselves, and the transmission model's emphasis is on moving messages around as an end unto itself. The latter activity is more easily quantifiable and commodifiable and much better suited to the marketplace and to industry.

It is also a cynical activity, insofar as it reduces values to numbers, by valuing only numbers. Others have noted this development by examining the substitution of marketing for collectivity, or, as David Marc's (1984) wry comment on Walt Whitman tells us, we are in an age of "demographic vistas." The result is a fueling of our distrust of the myth of progress and modernity, and fear that though we may never again be out of touch, we will rarely again feel touched by what someone communicates to us. That fear keeps us clinging to the communities within which we feel a sense of trust, of safety. In physical terms these are, increasingly, gated communities. In terms of computer-mediated communication these are "Gates-ed" communities, ones in which we hold keycards in the form of passwords, connectivity and access. In cyberspace these are what I believe is an analog of "metropolis": "Micropolis," namely, smaller and smaller groupings of people, fractal metropolii. I use the term "fractal" in this case both in the sense of a figure with self-similarity at all spatial scales, and as a play on words, a concatenation of "fractured" and "partial." Micropolis is a fragment, a fractured substitute in our lives for a polity. But it is also a fractal in the sense that social groupings in geographic, physical space, and ones in cyberspace, are gaining in self-similarity at and through all levels. Online, micropolii are gated in an oddly interlocking fashion [a gate opens into a community, but may also, like a cosmic wormhole, open into still another community seemingly very different and separate, though linked via interest (Jones 1995)]. Micropolii are, I believe, the result of what Marshall Berman (1982) identified as "The innate dynamism of the modern economy, and of the culture that grows from this economy, annihilat(ing) everything that it creates—physical environments, social institutions, metaphysical ideas, artistic visions, moral values—in order to create more, to go on endlessly creating the world anew" (288).

Interconnected though micropolii may be, they rarely form a collective via their interconnectivity, instead serving groups just slightly different one from the other. We experience a fragmentation of community just as we have on introduction and spread of cable television, magazines, and numerous other media. Our sense of others is very wide, our experience of others not very long. Perhaps this is due in some part to the approaching end of the millennium, a time when life seems to simultaneously speed up and slow down, the former feeling aroused by our sense of the length of time, the latter brought on by our sense (to borrow from Laurie Anderson's observations during her performances) of time's width. As we sit on the cusp of millennial change, we not only feel that time stretches very far back, that it has a retrograde trajectory, but that it stretches very far ahead, too, perhaps so far ahead that we cannot comprehend, and as we near the year 2000 the millenium becomes a handy marker for us, a time buoy if you will. It bobs along, always at a seemingly unchanging distance from now, though I wonder how that distance will affect us in 1999 when we can no longer use years a measure that keeps us distant from millenial change.

Perils and Parallels

A friend once remarked that "no one ever said that change had to make any kind of sense at all," a statement both true and revealing. Its truth is rooted in the randomness of change, in the inability to, god-like, will everything into place. It reveals that we nevertheless try to make sense of change, whether we try to will change into being or not. And perhaps we work even harder at sense-making as we become ever more sensitive to the ephemeral nature of meaning. The activity of sense-making has, in the case of life in compunities, made clear four areas that are common, forming a consistent narrative pattern illustrating where social concerns lie: privacy, property, protection, and privilege. That these themes are central to our discourse about new communication technologies is telling both because it makes our concerns clear and because it points out the mythic nature of technology's promise. The former is not difficult to discern, as these themes are easy to find in our conversations about the Internet and compunity. The latter is no more difficult to discern either, but requires the historicizing of these narrative patterns to help explain the role of new communication technology in social change.

Privacy

Much of the current discussion about the information superhighway revolves around privacy. It forms the core of many a government's concern that a "back-door" must be created for every computer and network (using the "Clipper chip" in the US, for instance) to allow access for the computer equivalent of continual surveillance and eavesdropping. In more commercial terms, one can ascertain corporate interests in gathering information electronically from us as well, and perhaps the most notable such attempt via computer-mediated communication was Microsoft's intention to include as part of its Windows 95 operating system a program element by which, upon electronically registering the software, information about a person's hardware is transmitted to Microsoft.

Privacy also forms the core of concerns about how information about ourselves will traverse the highway. Will anyone be able to "tap" into the data stream and fish out our credit or medical records? Will they be able to intercept credit card information as it zips from Internet site to Internet site? How will we prevent that from happening? What will happen to all the data that we send? Since data is relatively easy to store, will every message we send and receive find a place in some great universal archive? In place of gossip and hearsay, features of community, we find control and manipulation, features of compunity. These issues have followed the development of each new communication technology, from the advent of writing and printing, through the invention of television, when we thought others would see into our living rooms via the picture tube, and are symptomatic of a larger social issue, namely the ebb and flow of the boundary between public and private. To borrow from Walter Ong, what drives our concerns is the seeming permanence of methods of communication beyond the oral. As regards the spoken word, once something is uttered, it is also lost to all but memory, and as we have become less trusting of our own memory (illustrated by brisk sales of Dayrunners, personal organizers, etc.) we also become inversely more trusting of our ability to deny that which was once spoken as having been misheard, misrepresented, misinterpreted or simply incorrectly remembered.

In essence, our privacy concerns are based on the need for externalizing (or commodifying), in a more or less permanent fashion, information about ourselves. It too needs to travel, to be transported, and it needs to do so independently of us. We cannot be in more than one place at a time, but social relations, particularly ones formed and maintained by bureaucracies, demand that we be. And once information about us is external to us, it is also out of our control, just as the picture once taken of us is no longer ours but the photographer's.

It is important to note that one perspective on privacy issues runs parallel to what Jean Baudrillard (1983) has written in regard to the hyperreal, the "realization of a living satellite," in which "each person sees himself at the controls of a hypothetical machine, isolated in a position of perfect and remote sovereignty, at an infinite distance from his universe of origin." Our privacy is to a large degree not based on the need to control what is "inside" us already, but to control what escapes us and enters domains other than our own "private," and to conversely control that which does enter our own private sphere. Internet technologies are the electronic component (and a natural evolution of the telephone) to the triumvirate of technologies of the Fordist project of suburbanization. The first component was the development of the modern house, removed from the street, fenced off (and in some cases within gated communities) from others. The second component was the automobile that allowed movement along a physical network of roads and highways that managed to provide access to places outside the house while maintaining minimal contact with others. The metaphor of the Internet as "information highway" thus has another parallel, to Fordism, particularly as it engages Fordist notions of efficiency, supplanting a mechanical system with an electronic one.

But to control information to the extent that we can manage not only its movement from our own selves into the public realm but its subsequent metamorphosis in and during public discourse is nearly impossible, and denies that we are public beings, denies our essential humanity. We can no more control information, once externalized, than we can control the propagation of waves from a raindrop that has fallen into a pool of water. Of particular concern, then, is that continuing emphases on privacy concerns, by engaging us in a frenzy of largely unproductive activity to ensure that we control our inner and outer worlds, do, to some extent, more than symbolically privatize us more than we may want or need.

Property

Relatedly, once information about us is made external to us, and subsequently made digital and available electronically, its dissemination is relatively not complex. Copying files on disks or sending them over networks is electronically and mechanically much, much easier than photocopying a book, for instance.

But more interesting than simply the ease with which we can accomplish copying is that ultimately, given that information in the digital domain is essentially string upon string of ones and zeros, we are beginning to redefine the term, and perhaps very nature of, "property." Who owns a numeral or a "bit"? We have some evidence of the nature of that question from experience with software and compact audio discs. When we can not only copy but clone things, how will we identify "originals"? And, more importantly in industrial (and again, Fordist) terms, how will we restrict production and acquisition to effectively control the marketplace? Copyright law from its very beginnings relied on adjudication, not enforcement, by the government. For enforcement it relied on technology. In the past copying a book was labor-intensive, and the process itself mitigated against copyright infringement. It was simply easier to buy a book than to copy it. The photocopying machine changed that equation of time and money, just as the cassette deck changed the relation between consumption and copying for music, the VCR changed it for films and TV shows, and the computer changed it for software.

The most often asked question in this regard is: What will authors and publishers do to ensure income from their work if it's available on an electronic network? The issue is not in the first instance one of economics, but again one of control. Who will have the right to do something with a work is not a decision inherently connected to determining who will profit from it. As with aforementioned privacy issues, control is the root concern, for as soon as we have externalized (commodified) a work, it can migrate away from us in the same fashion that credit and medical (or any other) information can be passed around.

Moreover, control is the primary concern of entertainment and electronic industries that struggle with the structural overcapacity of production whose only traditional solution (one in name only, for each solution has begotten another problem) has been the evolution of distribution. Consequently, the development of distribution channels has outpaced the ability of the sociolegal complex to maintain a civil order that has traditionally offset the tension between publisher and author, the two sides of the production chain that coexist least easily. The Internet is thus a project alongside that of the opening of markets and borders, epitomized by the GATT and the NAFTA, trade agreements that provide the greatest freedom to movement of abstract commodities, or, namely, intellectual property. The development of the Internet has bumped up against legislative issues, and is only further evidence that the decentralization of distribution as an aid to mass production and consumption, is in fact inimical to control by legislative means.

Protection

If legislative means are unable to protect us from the flow of information, what might? To return to the concept of electromotive force, the lines of magnetic force created by a current flowing through a wire are directional, and move in the same direction as the current's flow. Moreover, these magnetic lines of force are elastic, and cannot be broken. One might imagine that the current is that which is created, distributed and consumed, and the magnetic force is the sociocultural change occurring external to such a Fordist system.

Historically, protection has been understood as the attempt to regulate the "current," in this case, namely, the content of what flows through the system. Consequently, authors have long sought protection for their work, but it has been producers, manufacturers, and distributors who seek ways to ensure income, and to do so requires some form of protection against copying. However, experience (particularly recently with Digital Audio Tape and its Serial Copy Management System) has shown that a technological anti-copying solution is rarely a final solution. For many authors the concern over copyright has as much or more to do with having their work re- or de-contextualized than it does with financial gain (the US is one of the few countries that does not recognize an author's moral rights in a work).

There is another way to think about protection vis-à-vis content, as that which protects the integrity of a work. The technology that enables both new forms of creative activity (desktop publishing, collaborative writing, computer-aided design, digital audio and video, for instance) also enables its distribution via new media like the Internet, and enables its ready editing and recombination. What, if anything, can protect the integrity of a work that new technologies make so malleable?

In fact the sociolegal system has had less difficulty with these issues than it is now having, and is going to have, with issues related to the "magnetic fields" (to return to the metaphor of electromotive force) created by content. To put it another way, the technologies of content distribution also deliver meaning to us. We will likely want to avoid some of it, we will want to screen some of it, and some of it we may, for good or ill, feel a need to censor. We will seek protection in the same way some now seek it from violence, obscenity, and the like found in older, traditional forms of media. We may also seek protection from the equivalent of "crank" phone calls, and from the inability to verify identity of the senders of messages. These are the concerns of legislation such as that found in portions of the telecommunications bill passed in the US in 1996. What such forms of legislation seek to protect against is not content *per se*, but the consequences of content. We sought (and continue to seek) such protection from the telephone, television, radio, telegraph and virtually all other media, for they are not merely "media" in any kind of passive sense, delivering information and nothing more: they are active intruders into our mental processes, requiring our attention, which, whether freely given or not, is not returned.

Thus it is, I believe, that we seek protection from what we have termed "information overload" (no matter how much, on some level, perhaps only the commercial, we may wish to be the ones doing the overloading). The question here is: How do we attend to the social connections impinging on us, the connections we at once desire (e-mail, telephone, fax, etc.) and despise (for they take up more and more of our time and energy)? These are the lines of force created by the "current flow" of content. We couldn't be more in touch and yet the telecommunication industry promises us ever closer, faster and greater contact. It is necessary to think through the implications for a society whose members face ever-greater demands on their time and thought. These demands make it more difficult than ever to engage with others by non-technological means, and shave away the time we allot to personal interaction. They are but one form of communication, perhaps neither better nor worse than any other, but they do carry with them their own structuring forces.

Privilege

Among the structuring forces is that of access and it will not be equal and uniform. To have it so would mean, in social terms for instance, not only provision of hardware and connectivity, but operating systems so sophisticated as to be stupid, that is, sophisticated enough to know when users are unsophisticated and then able to "dumb themselves down." It would mean the technological equivalent of "a chicken in every pot." It would mean the establishment of universal literacy, for, if nothing else, using computer networks requires good reading and writing skills. But, most importantly, it has already meant the definition of computing as a social necessity.

Will we have information "haves" and "have–nots"? Probably—we already do, with or without computers. What will be the consequences? That is more difficult to determine. We already have such a class separation—in some sense those reading this essay are likely to be "haves," and others, from different backgrounds, different experiences, different opportunities, may be destined to be "have nots." There are at least two important questions resulting. First, what will you do with what you have? Second, what will it be like to have it?

There is also the matter of privilege in its more mundane sense, and for those in education, publishing and related fields, this is critical to understand. Again, the latter sense of privilege is directly related to the initial lines of force created by the passage (movement, transportation) of content across new networks of communication. The more common sense of privilege I wish to invoke here is related to the lines of force created at right angles to that initial force, the "magnetic" instead of the "electrical" in terms of electromotion. We do not have information elites in the sense that the "haves" simply have more information than others, but in the sense that it is the "haves" that are organizing information for others, and by so doing they are undertaking a profoundly socio-epistemological act, generating the maps, indices, tables of contents, bibliographies, hypertext links, that others will use to organize not only their research and writing, but their thinking and knowledge as well. We have witnessed these past few years (at least) the eruption of critical scholarship that, for instance, critiques New World narratives and seeks to restore understanding of indigenous cultures and knowledge. May we be self-critical as we undertake an enterprise similar to that of New World explorers, who came, saw, and categorized?

Conclusion

It is by a very slow and gradual process that social change motivated by new technology, and new media technology in particular, occurs. We do not shift from one paradigm to another, from one process (mental or physical) to the next, at all quickly, and, I would argue, we often do not notice change when it does occur, because it does not happen in the expected social arena. So, for instance, the widespread use of the printing press and the spread of literacy lead to increased education and awareness, which we expect, but they also lead to isolation, which we expect less, even though we have greater awareness, for as we attend to our reading material we attend less to those around us at the time we are reading (which we often find useful when we sit next to strangers on an airplane, for example). Consequently, I am quite unsure about the potential to harness any technology for predictable social change. Our technologies are designed in anticipation of their effects, but the effects themselves are not ones that are informed by history, rather they are woven from our hopes. We seem to be taking a step toward privatization and polarization through use of new communication media like the Internet, but is that symptomatic, causal, or . . . ?

Irrespective of the answer to that question, we ultimately need to examine our assumptions about how new media technologies will affect our society. We seem to hold some common beliefs (Thornburg 1992), that they will:

benefit education and learning;

break down barriers and hierarchies (social and other kinds);

create new social formations, typically in opposition to dominant ones;

make participatory democracy feasible and easy;

make the interface between man and machine seamless; and

create new legal and ethical problems outside the parameters of existing policy and legislation.

Where do these assumptions originate? Have we tried to achieve these things already, by other means, and with what success? Or do they remain assumptions (or hopes), realizable or not? Our ethics must spring from our beliefs, and as yet our beliefs about technology are uncertain, just as the technologies we envision are not certain, and indeed are consistently in flux. But we do not need the technology to look inside ourselves, we need only to inspect our beliefs and reflect on them, for they, and not the technology, represent what we desire.

Other outcomes are just as possible, and to an extent are already making themselves present. Our use of an index, for instance, is being replaced by a point/click/search paradigm establishing itself through use of hypertext, electronic databases, the World Wide Web, and the like. In education the busywork that teachers once handed out via paper is often being supplanted by busywork via computer and touted as somehow more beneficial to students on account of its "interactivity," though in such cases interaction is so loosely defined as to mean anything from pushing a button on a mouse to attending to an audiovisual presentation. These are outcomes, to use the concept of electromotive force a final time, at "right angles" to the ones most visible. They affect our everyday lives in innumerable ways, remain elastic but not breakable, affect our thinking and very thought processes, but do not come at us in one fell swoop, and are often difficult to describe, much less to desire.

It is particularly important to note that, on reflection, each of the above beliefs is rooted in the transportation model of communication, which is itself based on the primacy of the movement of current through a wire and unreflective of the "right-angled" lines of force. Each belief in its way has as its premise that moving messages around more effectively will make these beliefs metamorphose to reality. Perhaps this is not surprising, for in Western societies, to a great extent, transportation has been a ritual activity. Unlike in our public social lives, in many ways one of the few activities over which we have a great deal of control is transportation. Our own bodily "technology" evolved toward mobility, and we have used technology to augment it. We are at the wheel of our car, our control panels in front of us, regulating our own private environment. And cars and driving are not the only area in which we increase control of transportation—we effectively increase it via the new technologies of communication, by using fax machines and e-mail, time- and date-stamping messages, and packages and memos, ensuring that our words and information get where we want them to go, and do so on time, through a variety of control mechanisms. In fact, one of the most touted aspects of the combination of telecommunication and computers is that it will somehow supplant transportation altogether and result in a great increase in telecommuting. That, so far, has not happened, but it presents an interesting, and heady, mix of metaphors that have driven (pardon the pun?) national conversations in Western countries, and continue to fire the futurist manifestos of many politicians, particularly ones in the US Congress (as well as marketing pundits).

We still lack control over what will happen to the messages we create and send when they get where they are going, because they are essentially out of (our) control. I do not believe any form of technology can assist us to better create and interpret messages—only we ourselves have the capacity to better those abilities. It is most disheartening, perhaps dangerous, to believe that since machines have replaced some forms of human labor they will replace human thought. Perhaps the greatest force mitigating against telecommuting, and ultimately against most technology, is that people like people, seek to be with other people, and seek to maximize interaction. Developers of tools like those associated with the Internet's use succeed best, it seems, when they recognize that, and put technology in service of conversation rather than communication, in service of connection between people rather than connection between machines, and in service of understanding rather than movement.

Note

This manuscript appeared originally in the *Electronic Journal of Communication / La revue electronique de communication*, 8 (3 & 4), 1998 (see <http://www.cios.org/www/ejcrec2.htm>) and is reprinted by kind permission of the editors.

References

Baudrillard, J. 1983. "The Ecstasy of Communication." In *The Anti-Aesthetic*, ed. H. Foster, 128. Port Townsend, WA: Bay Press.

Berman, M. 1982. *All That is Solid Melts into Air*. New York: Simon and Schuster.

Carey, J. 1989. *Communication as Culture*. Boston: Unwin-Hyman.

Cheney, M. 1981. *Tesla, Man Out of Time*. Englewood Cliffs, NJ: Prentice-Hall.

Deleuze, G., and F. Guattari. 1980. *Milles plateaux*. Paris: Minuit.

Eisenstein, E. 1979. *The Printing Press as an Agent of Change*. New York: Cambridge University Press.

Forster, E. M. 1948. *Howard's End*. New York: A. A. Knopf.

Foucault, M. 1977. *Discipline and Punish: The Birth of the Prison*. Trans. A. Sheridan. New York: Pantheon.

Goody, J. 1986. *The Logic of Writing and the Organization of Society*. New York: Cambridge University Press.

Jones, S. 1995. *CyberSociety: Computer-Mediated Communication and Community*. Newbury Park, CA: Sage Publications.

Marc, D. 1984. *Demographic Vistas*. Philadelphia: University of Pennsylvania Press.

Marvin, C. 1988. *When Old Technologies Were New*. Oxford: Oxford University Press.

McLuhan, M. 1965. *The Gutenberg Galaxy: The Making of Typographic Man*. Toronto: University of Toronto Press.

Ong, W. 1982. *Orality and Literacy*. London: Methuen.

Patton, P. 1986. *Open Road*. New York: Simon and Schuster.

Thornburg, D. 1992. *Edutrends 2010: Restructuring, Technology and the Future of Education*. Mountain View, CA: Starsong Publications.

Electronic Networks and Civil Society: Reflections on Structural Changes in the Public Sphere

Barbara Becker and Josef Wehner

Introduction

The contemporary media system is undergoing a rapid and fundamental change caused by the emergence of a new electronic medium: the "Internet." In comparison with mass media, this new medium offers relatively cheap and simple access to worldwide information and communication opportunities. It supports multilateral communication without the barriers of natural interactive communication (above all, dependency on the presence of the participants, and lack of time to meet face to face). Compared to mass media and their journalistic professionalism, people do not need much money or any special higher qualification to use the Internet. They can publish their points of view, ideas, and comments to a particular topic without being restricted by time and space, and without depending on greater organizational or professional support. In this way it becomes relatively easy for every participant to act as an editor or a publisher. So, the revolutionary significance of the Internet is caused by an increasing independence from the traditional "intermedia" system (mass media, political parties and other representative institutions).

Therefore it is no surprise that people have high expectations about the societal consequences of the Internet with respect to the public sphere. Looking at the current debate on this theme, we would like to comment on a polarization between two different positions (see Leggewie and Maar 1998). One side emphasizes the democratic potential of the Internet. According to this optimistic interpretation, the Internet may strengthen the position of citizens in relation to political authorities by allowing them to participate directly in political decision-making (Grossman 1995). This view emphasizes free access to information, globalization and lack of censorship as central characteristics of the Internet. Accordingly, electronic communication may introduce a new form of direct or plebiscitarian democracy because it seems to introduce new forms of participation, equality and social bonding. As a result, traditional mass media will be replaced by electronic networks. The new media, by allowing direct contact between citizens and political representatives, will lead to a new form of the public sphere which is no longer controlled by a few actors. This structural change of the public sphere supports the preference for individual and collective autonomy and political decentralization. From this perspective, the process of technological innovation corresponds to processes of differentiation between individuals, milieus, and kinds of communities in the modern world by providing them with the possibility of articulating their interests and opinions autonomously.

The more skeptical position maintains that the vision of the Internet as a free, anarchistic medium belongs to the past. In the meantime the free and uncontrolled communication structures that were the main characteristic of the Internet in its early days are more and more confronted by the development of new hierarchies, new centralized structures, and new ways of protecting, controlling, and concealing information. It is feared that the Internet is no longer a free and accessible communication environment (Maresch 1997). A tendency can be observed which shows a dichotomy between public and private channels on the Internet. The public domain is accessible for every user, but only takes up rather trivial information, while private channels, which are politically more relevant, are not accessible to everybody but only to a small group of responsible and powerful people. Thus, we find a tendency towards establishing centralized structures of demarcation and exclusion.

Our position lies in between these two points of view. Referring to both theoretical and empirical investigations, we will show that the public space based on electronic networks is something qualitatively different from an all-inclusive public based on the mass media. The constitution of an all-inclusive public depends on a type of media distributing texts and pictures which can reach recipients at the same time and in an identical form. In contrast, electronic networks provide individual experience and heterogeneous opinions. They constitute a multitude of simultaneous partial publics grounded on a broad spectrum of issues. So, we can perceive the Internet as an independent, separate type of medium, the effects of which are emerging in other than mass-media related areas. Electronic networks will certainly change political public opinion, but in ways that are supplementary to existing forms of public communication. In particular, we think that by virtue of its interactive communication structure, the Internet may support the domain of public communication, which has been described as "civil society" in the context of theoretical discussion about modern democracy. The term "civil society" refers to a network of pre-institutional civil activities and assemblies as well as social movements and pressure groups (compare Seligman 1992). These movements form an alternative public sphere, which influences both political decisions and the public opinion established by mass media system. In this way civil society generates partial forms of public opinion which are relatively open, close to the needs of citizens and which are characterized by rather elaborate levels of discussion.

Media and the Public

Modern societies can be described as functionally differentiated societies (Luhmann 1997). Therefore one may say that meaningful structures, which are still connected to each other in premodern societies, have been separated from each other and independent structures and functional systems have emerged. The differentiation of society has, however, not only improved its efficiency, but it has also generated problems. The autonomy and closed character of the different functional systems imply opacity between systems because of the high inner complexity of each system; moreover, we observe that systems ignore possible consequences of their own behavior regarding other systems. As a result, modern societies are always confronted with problems of integration, unification and self-description.

Here, the political system manifests some particularities (compare Luhmann 1971). On the one hand, as an autonomous system, it has the same characteristics and problems as any other functional system. But on the other hand, the political system has to find methods of integrating different perspectives in order to guarantee the consistency of a society. The political system has to cluster the different perspectives of the functional systems within a society to generate a transcontextual consensus. And, in modern democracies, only the political system is seen as the legitimate system for formulating general rules and frameworks for other systems of society.

In looking for institutions that enable a society to find this kind of transcontextual agreement, public communication structures seem to play a fundamental role. On the one hand, they comment on political decisions and explain them to the citizens; on the other hand, they collect expectations and demands of the citizens and present them to the political system. Political public opinion therefore may be regarded as a communication system that mediates between the citizens and the political system by reciprocally selecting and transferring information (see Jarren 1998). From the past we can distinguish between three different forms of public opinion: the public sphere grounded on encounters, the public sphere based on assemblies, and the public opinion generated by mass media (Gerhards and Neidhardt 1990). Through specific ways of selecting, clustering, and disseminating information, each kind of these public spaces involves opacities as well as discoveries of new forms of perceiving and constructing reality. Through particular ways of selecting and codification media generate specific ways of world-making as well as areas of blindness. In the context of this paper, we would like to profile the particularity of public opinion based on electronic networks in relation to the public space based on mass media.

Mass Media and the All-Inclusive Public

Nowadays it is well known that modern societies generate numerous orientation problems. The permanently growing abundance of information and communication possibilities cannot be coped with without the application of technical media as reduction mechanisms. It is the function of newspapers, radio, and television to reduce the complexity of these channels of communication—not only for a special group, but for the whole society—down to an accessible scale (see Luhmann 1996). Mass media does this by providing messages which ". . . have an intrinsically public character, in the sense that they are 'open' or 'available' to the public" (Thompson 1995, 31).

This requires not only special techniques of filtering and preparing information, but, both in a technical as well as in a organizational sense, a structured interruption between the production of information and their reception. The public, generated by mass media, is not based upon the direct participation of citizens but on centralized ways of selecting information and focussing on specific topics. Mass communication offers no possibilities of a direct feedback. It is based on different roles known as "sender" and "receiver." The messages must be transmitted from the producers to the recipients. Thus, the context of production is structurally decoupled from the context of reception. The situation in which the relevant events are gathered and the messages are produced is different from the situation in which the message is received. Producers and recipients are unequal partners in this process of symbolic exchange. The recipients have relatively little power to exercise influence on the topics of communication. In addition to this, the access to mass media and the possibility of rising to speak in mass media are not egalitarian—that is to say, relatively few people can influence this kind of public opinion. So, institutions and organized actors like firms, political parties, etc., face fewer obstacles to presenting their opinion in mass media than single and non-organized actors. Compared with such institutions and organized actors, the influence of the recipients is restricted to more indirect ways, such as readers' letters, etc.

But only these organizational and technical conditions enable a statement, a sound, or a picture to be infinitely reproduced, and therefore distributed in an identical form to, in principle, an unlimited number of recipients—manifested in the distribution of live events watched by hundreds of millions people worldwide. By watching television or reading a newspaper people get the impression that they are receiving the same information at the same time as an unlimited number of others. Thus, mass media draws the attention of an unlimited number of recipients towards a limited number of topics and statements.

The mass media guidelines of visualization and textualization create the conditions that allow the most extraordinary events in the remotest parts of the world to be translated into an anonymous sign system, through which they become accessible to a world wide distributed audience. These standards have, in addition, contributed to the development of international arenas in which the representatives of different national institutions (for example the spokespersons of different governments) react to, and are able to communicate with, one another by specified rules of conduct and rhetoric on events of world political significance. After all, mass media support the distribution of global economic, political and cultural standards, and provide a field of comparison for the relative national or regional variations.

In this way a public information world emerges that can reach transcontinental dimensions in this age of satellite communication. Following Niklas Luhmann, mass media generates "a background knowledge which provides a starting point for communication" (Luhmann 1996, 121). They can be compared with a great "mirror" in which everyone can observe what the others observe. By this, the reception of mass media products constitutes a (world) public arena including the whole spectrum of social collectivities, in which communication situations are pre-structured by themes for which "universal acceptance can be assumed" (Luhmann 1996, 22). Without this mirror the (world) society would not be observable or communicable as a "world society," and therefore not capable of reproduction. It would, instead, disperse into fragmented areas and cultures which would not share the messages selected by mass media as a common background.

The term "mass society" can therefore be understood as a global spread of messages and code patterns created by mass media. These are applied to every recipient, in principle, and therefore to a mass public, but leave open how they are interpreted. Mass media create, according to Featherstone (1990), a sort of "meta-culture" because of the cognitive schemata they bring into circulation, and therefore a collection of codes of perceiving and constructing reality, codes available world-wide. The messages and views of life spread by mass media cannot be translated completely into the multitude of constituent social contexts; they can, however, be adapted according to specific preferences and interests. In this way symbolic forms are created which cross all traditional forms of socialization and class. So, the public of mass media serves as a reference point for an unlimited audience and for social distinctions as well.[1]

Participation in a common information world depends on the selection monopoly of mass media, which allows for no interference from the recipients. It shatters as soon as the technically- and organizationally-based role definitions between sender and receiver are removed (compare Wehner 1997). The inclusion and synchronization services of mass media further depend on the fact that their products are presented in such a way that the communicator remains invisible (compare Schmidt 1994). If mass media were to show themselves as communicators, their texts and pictures would be recognized as selective interpretations of the world, and would no longer be accepted as an objective view of current events. Television is only a "window to the world" so long as the viewers ignore the mediating function of the medium and therefore succumb to the fiction of directly experiencing the events presented in pictures and text. It is only under these conditions that the instruments of mass media, such as television, represent more than just a further possible observer's perspective among many others. And only in this way the recipient does allow this to become a common backdrop used for the proliferation and comparison of personal perspectives and points of view. Therefore, mass communication has to remain impersonal; its themes have to address an anonymous mass public. From this perspective, strategies which try to involve the receivers as communicators in mass media events make little sense. They may open up the possibility of a personal arrangement of media products but then this is inevitably cancelled out at a later stage by the distribution methods of the mass media, which are based on anonymous communicator mechanisms.

From an All-Inclusive Public to Partial Publics

In contrast with mass media, electronic networks have been described as "individual media" which are used as a forum of nonestablished personal opinions and discussions (compare, for example, Rheingold 1994). By abolishing the communication and interaction barrier between sender and receiver, electronic networks offer an unlimited number of participants the opportunity to act as communicators, and thereby circumvent the anonymity and one-sidedness of mass media's production of meaning. In the Internet there are neither criteria for preselecting information, nor an efficient control of themes (except on the level of particular newsgroups or mailing lists). While mass media restrict the user to an "exit option" (Hirschmann 1974), the user of electronic media has also a "voice option"—that is to say, he or she can articulate his or her own opinion about a selected theme. The difference between the "balcony" and the "stage"—to say it metaphorically—which characterizes mass media, does not exist in electronic networks because everybody is engaged in the process of producing and spreading information. It is not only possible to get information from the Net, but furthermore, people may participate in chat groups and present personal viewpoints and particular arguments for discussion. Thus, participants are able to change between the role of an active communicator and a passive consumer. If they want to discuss a topic they may present their ideas suggesting a discussion about them. *A priori*, nobody and no topic are excluded. From a more abstract point of view we can argue that in electronic networks the context of producing a message is no longer separated from the context of its reception. Its interactive structure helps to overcome the traditional dual role of producer and consumer. Accordingly, the Internet shows a high diversity and plurality of themes.

Above all, the Internet is attractive because it presents the opportunity to produce thematically concentrated and specific communication relationships (compare Wellman et al. 1996). Messages on the network appear to be applicable to personal or group-related interests or activities. Looking at processes on the Internet, we seldom find discussions about global perspectives and topics which are relevant to the society in general. Rather, participants concentrate on very particular themes and more private needs (see Helmers et al. 1998). It is worth noting the results of empirical studies according to which electronic networks tend to segment the flow of communication (for example in mailing lists, newsgroups, and chat rounds). As it is difficult to find transcontextual perspectives and themes, most of the participants are limited to their specific discourses and do not even try to open them up to a more general discourse. In addition, electronic contacts are dependent upon a minimum amount of the users' assumptions and interests (compare Baym 1995; Jones 1995). In comparison with a mass public, the participants in an on-line public need to have specific insider knowledge and often, they show a relatively high level of homogeneity of their interests and abilities. If electronic networks already support existing structures and activities, they also reach a high level of organization (compare Wellman 1997).

In addition to this, everybody can present his or her points of view in his or her own language, which is separate from the censored and generalized code which regulates the way in which information should be presented in mass media (and is presented by journalists). Accordingly, the Internet shows not only an overwhelming multitude of topics but also a high diversity of special codes of expression. For example, participants in so-called virtual communities demarcate their activities from other groups by elaborating idiosyncratic styles of communication and specific language codes (Becker and Mark 1999; Poster 1995). They use the Internet as a public space for articulating their differences. First of all, "THE citizen" who converses with other citizens on the Internet does not exist. Rather, there are the representatives of special organizations, groups or social milieus—such as experts, old people, homosexuals, women, men, children, youngsters—who talk about their particular interests on the Internet.

Hence, electronic networks underline the internal differentiation of society by generating polycontextural communication structures. This new media technology thus fits in with current trends of further segmentation of society, because it strengthens the dissemination of pluralistic and incommensurable discourses instead of supporting the generation of commonly shared beliefs. So, the electronic communication in the Internet may be regarded as an example of a postmodern culture of communication. Paradoxically, the demands for equality in the world of communication promote the formation of global communication systems with exclusion criteria, so that no common body of networked communication can be formed. Thus, the public that is produced through the Internet service is always particular, in spite of its global range (compare Fassler 1996, 440). Networks offer neither any centralized assistance or criteria of selection, nor do they have any sort of general thematic references, as seen in mass media: rather they open up a communication space with a multitude of decentralized selection channels of equal status, whose products are only destined to reach a special public.

The lack of economic and other regulative conditions often is regarded as one of the main reasons why the global communication society based on the new media has not already taken place. However, such conditions may only affect the diffusion and the scope of the new information technologies. Differences in the information habits and attitudes towards different sorts of media are also not a satisfactory explanation as to why the hope of an unlimited inclusive communications society has, up until now, remained unfulfilled. Thus, the very nature of the media and of its social function should be taken into consideration. From this point of view it has to be assumed that new electronic media should not be expected to constitute a new mass media. Instead it is much more likely that they constitute and support partial public arenas. Such partial public arenas can be defined as social networks of users mutually communicating and informing one another on a particular theme ("special issue").

Electronic networks provide a forum for opinions beyond the officialdom of public reportage as distributed by the mass media (Aycock and Buchignani 1995). Thus, Internet communication is more characterized by presenting personal opinions, rumors, and individual comments on events which have been previously reported in the mass media. Mass media tend to suppress the significance of local background conditions and specific contexts. Communication on the Internet, however, tends to encourage these subdivisions by catering to the explicit preferences and interests of their subscribers. Public arenas produced by electronic networks are only partially public arenas. They open up a communication space between the level of the public as produced by mass media, and the level of public arenas which are constituted solely from the medium of the normal language in the form of encounters and assemblies. This intermediate public is in need of new communication technologies, because it is often characterized by a global spread of its communications on the one hand, and by a strict focusing on selected themes on the other hand.

Certainly, this interpretation of the Net is confronted by some critical arguments. So, it is possible to argue that even in industrial nations, only a very small part of the population has access to the Internet. For many people, the personal computer and related equipment which would enable them to participate in on-line discussions are still too expensive. And in countries of the so-called third world, the situation is even worse: most of the inhabitants of these countries do not even have a telephone connection. The possibility of participating in the public space of the Internet therefore is restricted to a very small, prosperous part of the population. Another problem appears if we consider that not only financial resources are a necessary precondition of participation; in addition, people should be endowed, to use a term of Pierrc Bourdieu (1987), with "cultural capital." Most Net-users not only possess sufficient money and technical competence, but also a specific educational and cultural background which includes competencies such as speaking English and being able to present arguments in a rational way (see Wetzstein 1995). An active participation in on-line discussion groups presupposes the ability to overcome one's shyness and to articulate one's own opinion, to present arguments, and to deal with the anonymity of the communication situation (especially the fact that very often, a response to what one has posted may not be forthcoming). The habit of the passive consumer, into which people have been socialized for such a long time, cannot be broken very quickly. The transition from the role of the passive recipient to an active user of interactive media is still in its early stages. Researchers are also more and more afraid that the Net will become fully commercialized on the long run.

Some of these problems are not technical but "only" social problems. So, they could be overcome through appropriate social reforms and political support (see Schmid 1997). There is also the fact that the computer equipment which is needed for participation on the Internet is becoming ever less expensive. Consequently, more and more people will get an access to the new media. Further, more and more people outside of the industrialized world as well will learn the basic technical and communicative skills to use the new media. And, last but not least, there is some hope that commercialization will not dominate all spaces of the Internet. [The economic actors cannot transform the Internet completely into a market because this media is open for different modes of treating and coding. Neither one type of actor nor a single social system can turn the new media into a medium focussing exclusively on special values and procedures. There are always representatives of other social spheres like political parties or social movements who use the internet to extend their facilities. So, while computer networks will be used for electronic commerce or be commercialized by firms, there will be other electronic networks characterized by non-profit motives and values (see Kleinsteuber 1996).]

Electronic Networks and Civil Society

The functional differentiation of modern societies implies a plurality and multiplicity of different perspectives and ways of world-making. Therefore, modern societies can be compared to a chaotic field. Every partial system not only has to eliminate its internal complexity, but also determine its relationship to its environment and define its limits (Luhmann 1997). This is especially true for the political system as it is confronted with manifold expectations as developed by other functional systems. To eliminate this complexity, political systems have to develop strategies for selecting themes and topics. Considering this problem, public opinion can be regarded as a mechanism for preselecting relevant themes and reducing complexity by generating particular techniques to filter information.

In particular, mass media have undertaken this function and has even institutionalized it (compare Alexander 1988).[2] Journalists select and prepare information and themes by following particular criteria that prevent them from presenting everything as they focus on particular topics. So, television and newspapers, aiming at a mainstream audience, report on political events by taking the position of a gatekeeper and controlling what will be presented and what will be hidden. Processes of selecting and filtering information and presenting relevant topics to political institutions can be regarded as the most important function of mass media (Luhmann 1971). Only the conventional mass media present converging views about the expectations and needs of citizens. They focus the attention of different individuals and groups on a single issue and create a strong "public opinion" which can influence the attention of the politicians and the direction of the political decision making. By looking at the headlines of the newspapers and the news of the television magazines every political actor can see what political action has been considered as relevant and what effects on other political actors can be observed. The specific power of mass media is due to its capacity to motivate political actors to become interested in specific themes. Accordingly, the messages of mass media have to be formulated in a rather uniform language, or, as Gerhards and Neidhardt (1990) put it, public communication is the communication of lay people. It is neither a communication of experts nor private communication, because everything has to be presented in a way that the standard citizen can understand it. In this way, mass media are able to mediate between politics and citizens. Jürgen Habermas (1992) refers to all these characteristics as the "mobilizing function" of mass media.

By contrast, the Internet opposes the consensus-building system of mass media. While the mass media distribute an identical information set to different people, providing widespread common experiences and homogenizing opinions, the computer media opens a public space in which different people and groups express their idiosyncratic points of view. So, mass media constitutes a homogenized audience while the Internet gives rise to a multitude of different partial publics. There is a great plurality of on-line communities and groupings on the Internet which causes a fragmented public space with a multitude of special issues at the same time, so that there is no focussed public debate on a single issue and no unified public opinion. The Internet may be regarded as a kind of stage for particular interests and identities, but it is no medium in which to express and develop global political strategies and negotiate common standards for talking together (Buchstein 1996). On the Internet, it is rather difficult to find a kind of common language that would enable all participants to focus on central issues. Accordingly, it becomes more and more difficult to find general topics and a commonly shared foundation of perspectives and beliefs. A clustering of viewpoints and strategies is seldom found on the Internet and general objectives are replaced more and more by individual or group interests. The rhizome-like structure of the Internet and the complexity of links of the World Wide Web (WWW) allows the user to follow his or her own interests but these characteristics do not support the development of binding insights, gathering views, binding problem-solving strategies and common perspectives; rather, they support particular orientations and the differentiation of communication processes. It seems that the new media opens a room for a multitude of partial publics based on a broad spectrum of special issues. The heterogeneity of milieus and cultures, the differences between all the political movements and initiatives become visible and are supported.[3]

The multitude and heterogeneity of information providers attract the attention of the user to possible motives and interests of the sender, while the homogeneity of presentation in mass media suggests that there are no alternatives of topics and modes of presentation. But normally an Internet message cannot reach an unlimited number of people in the same form and at the same time. Therefore users never know if they have reached a mass public or not. Hence, electronic networks cannot replace the mass media's role as supporting specific kinds of selection and inclusion. Computer networks even seem to threaten a public based on an orientation towards generalized products of meaning. All participants can become active providers, but this inevitably leads to "parceling out" of the electronic space. Through networking, more and more participants have a voice; but because of the increasing number of participants there is less and less time to listen. This problem can only be solved by a new asymmetry of speaker and listener roles, or through a limitation of the communication configurations (Helmers et al 1998). Therefore electronic networks do not offer a functional equivalent to the mass media.

So far, the Internet lacks the ability to dramatize problems in a way that makes political systems take notice of them. There is no strategy for clustering different perspectives and discourses so that they may represent transcontextual themes and perspectives that could influence the process of political decision-making. So, at least for the moment, mass media cannot be replaced by electronic communication networks, as only mass media can guarantee this kind of transcontextual clustering of topics and is able to force political reactions. But we assume that the Internet increases the opportunities available to citizens for expressing their interests. Considering this, the Internet will influence political public opinion to a large extent, because new domains of discussion and new discourse forms will enlarge current ways of generating public opinion.

At this point, it seems appropriate to refer again to Habermas (1992, 435), who distinguishes between a kind of general public opinion generated by mass media, and a different form of partial public opinion which is less formal. Partial public spheres are characterized by variable non-governmental and non-economical associations and assemblies, (i.e. community pressure groups, political associations, etc.). In comparison with other political actors such as political parties, these grassroots movements are concentrated on specific issues; they are timely, restricted and the ties between their members are relatively weak. Together they constitute the so-called civil society (compare with Cohen and Areto 1992; Frankenberg 1996; Hall 1995) Each of these pre-institutional associations creates a specific public sphere. Debates on this level of articulating and defining political issues often contradict the general public opinion produced by mass media. These partial publics therefore can be understood as an important space of resonance for the "real" problems and interests of citizens. Mass media cannot exist without this foundation of deliberate associations, because otherwise, the formal structure and the clear professional separation between the producers and the auditorium would not be able to mediate between politics and citizens. Partial public spheres, like these non-formal associations, are more characterized by authenticity, creativity and sensibility; that is to say, partial public opinion is more open towards those problems and interests which are not represented in public opinion generated by mass media.

According to Habermas, partial public spheres have to intervene between the generalized and formal functional systems, on the one hand, and the vivid lifeworld of citizens, on the other hand. Social movements and community pressure groups raise the real problems of citizens and present them to political institutions and their representatives. Public opinion produced by such informal groups is not only a very important indicator of whether the system as a whole is functioning; furthermore, here exists a vivid power of innovation, critique and creativity through presenting different perspectives to political institutions and through considering the diversity of various viewpoints. In this respect, it seems that the new electronic media support the actual trend of citizens' political interests and engagement shifting from the traditional political institutions and their programs to more thematically focused and timely, restricted initiatives and movements without being forced to develop a deeper commitment ("mouse-click activism").

As we have already described, only organized actors such as political parties or powerful companies get the opportunity to speak to an audience with the help of mass media. The more mass media become commercialized and controlled by only a few large-scale actors like Time Warner, Bertelsmann, or Rupert Murdoch, the less individuals, groups and smaller-scale political actors have the chance to articulate autonomously their opinions in newspapers, radio, or television. Thus, on this level mass media tend to deny the empirical heterogenity of cultures existing in the society. Until now there has been only little technical support to enable unorganized people to publish their opinions. Apart from meetings and assemblies there were only small-scale newspapers or pamphlets to present their opinions and to organize their activities. The Internet has changed this situation by providing this zone of political communication with new technical capacities of public communication.

Through the Internet, individuals are provided with a broader range of insider information about news that has been disseminated by mass media. Anyone who has been forced to take a passive consumer position in the era of broadcasting can now make his comment on political events. People can articulate their views and are in the position to generate a new foundation of political and public opinion, in contrast with the more centralized orientation of old media. Because of the cheap access to the new media and the relatively low infrastructural and organizational requirements, even extremely small groups have the chance to present themselves independently in a public sphere. Above all, persons and groups who are affected by wars or other catastrophes, as well as members of marginalized minorities which had no voice in the public sphere, will profit from the new medium. By supporting these people and their preferences, electronic networks mirror the multitude of life conditions in various social milieus of the (world) society (see Geser 1996). Some examples, like the accident in Chernobyl, the Gulf war or the civil war in the former Yugoslavia, are indicators of such a form of an alternative public sphere. Environmental movements such as the World Wildlife Fund (WWF) or Greenpeace are examples of how social networks are established apart from traditional differentiation (milieu or party membership) alliances. At the same time—apart from traditional polarizations such as class conflict—such social movements create new counter-positions and new social conflicts. Electronic networks facilitate the emergence and formation of non- or trans-territorial solidarities and the process of networking between different movements and initiatives. They support the internal communication of social movements and associations and their related public arenas.

According to this, we may say that the Internet is much more capable of considering the various demands and diverse interests of individual citizens than the centralized, organized mass media. The Internet bridges the gap between the extremely selectivity of the mass media and the great variety of the private milieus and individual lifeworlds (see Ess 1996). It may strengthen, to use another special term of Habermas (1992), the “signal function” of public communication.

Conclusions

It is now obvious that electronic networks are mainly relevant in the field of politics. Enthusiasts see the chance to reform the representative constitution of modern democracies fundamentally in the new medium. From this point of view, the thematic filter and exclusive cast list of speaker roles in mass media can be overcome. In contrast with this, we assume that electronic communication networks support political partial public areas. In particular, nongovernmental organizations, social movements, and citizens' initiatives, amongst others, seem to be strengthened by electronic communication. The Internet will not replace the public space based on mass media. Instead, it constitutes a public characterized by pluralistic perspectives, an unlimited number of on-line discussion groups, and multiple connections between different "communication arenas." Because of its interactive and (until now) rather decentralized structure, the Internet enables people to present their perspectives and positions in a still rather unlimited way. So, we may expect that the Internet opens a new public sphere which helps to overcome some inequalities based on different possibilities of getting a voice in the public. But we have to consider that computer nets are characterized by their own specific potentialities and shortcomings. Considering this, we assume that the media world is progressing towards a higher degree of differentiation. Thus, the need for conventional mass media and their capacities to influence the public opinion will not vanish. But they have to specialize themselves more and more on those tasks which—at least in the moment—new media cannot replace.

Notes

1. The possibility of selecting and interpreting the common background information which has been distributed by mass media significantly contributes to the identity formation of exclusive groups or special cultures. The inclusion mechanism of the mass media and its accompanying processes of regulating attention are therefore in no way contradictory to the observable recontextualization of mass-media-mediated presentations of reality. In this respect the mass media supports processes of individualization as well as the differentiation of partial systems, all of which are typical of modern society. Mass media help people to find out from what, whom, and in what respect they differ. By relating themselves to mass media, people as well as groups are able to develop a distinctive identity (see Thompson 1995).

2. Whereas it is often assumed that this specification of various media also includes the mass media, we support the theory that in the future there will still be the need for media with an anonymous relationship to the public—possibly for the very reason that there is such a dramatic

growth in types of media and media products. Of course one cannot assume from this need that the mass media are guaranteed to survive in their traditional form. Rather, a problem can be defined: if the mass media were to be transformed into specialized media for insiders, this would mean that society would lose the possibility of self-observation and description acceptable to all areas and cultures of modern society.

3. This is one of the main reasons why people often complain about the lack of order and orientation on the Internet.

References

Alexander, J. 1988. "The Mass News Media in Systemic, Historical and Comparative Perspective." In *Action and First Environments: Towards a New Synthesis*, ed. J. C. Alexander, 107–52. New York: Columbia University Press.

Aycock, A., and N. Buchignani. 1995. "The E-Mail Murders: Reflections on 'Dead' Letters." In *CyberSociety: Computer-Mediated Communication and Community*, ed. S. G. Jones, 184–232. Thousand Oaks: Sage.

Baym, N. K. 1995. "The Emergence of Community in Computer-Mediated Communication." In *CyberSociety: Computer-Mediated Communication and Community*, ed. S. G. Jones, 138–63. Thousand Oaks: Sage.

Becker, B., and G. Mark. 1999. "Constructing Social Systems Through Computer-Mediated Communication." *Virtual Reality* 4: 50–73.

Bourdieu, P. 1987. *Die feinen Unterschiede*. Frankfurt am Main: Suhrkamp Verlag.

Buchstein, H. 1996. "Cyberbürger und Demokratietheorie." *Deutsche Zeitschrift für Philosophie* 44: 583–607.

Cohen, J. L., and A. Arato. 1992. *Civil Society and Political Theory*. Cambridge, MA: MIT Press.

Ess, Charles, ed. 1996. *Philosophical Perspectives on Computer-Mediated Communication*. Albany: State University of New York Press.

Fassler, M. 1996. *Mediale Interaktion: Speicher Individualität Öffentlichkeit*. München: Fink.

Featherstone, F. 1990. "Global Culture: An Introduction." *Theory, Culture and Society* 7: 1–14.

Frankenberg, G. 1996. *Die Verfassung der Republik: Autorität und Solidarität in der Zivilgesellschaft*. Baden-Baden: Nomos-Verlag.

Gerhards, J., and F. Neidhardt. 1990. "Strukturen und Funktionen moderner Öffentlichkeit: Fragestellungen und Ansätze." WZB [Wissenschaftszentrum Berlin] Discussion Paper FSS III 90–101. Berlin: WZB.

Geser, H. 1996. "Das Internet—ein Medium 'herrschaftsfreier' politischer Kommunikation?" In *Politisches Raisonnement in der Informationsgesellschaft*, eds. K. Imhof and P. Schulz, 213–27. Zürich: Seismo Verlag.

Grossman, Lawrence K. 1995. *The Electronic Republic*. New York: Viking, Penguin.

Habermas, J. 1992. *Faktizität und Geltung: Beiträge zur Diskurstheorie des Rechts und des demokratischen Rechtsstaats*. Frankfurt am Main: Suhrkamp Verlag.

Hall, A., ed. 1995. *Civil Society: Theory, History, Comparison*. Cambridge: Polity Press.

Helmers, S., U. Hoffmann, and J. Hoffmann. 1998. *Internet . . . The Final Frontier: Eine Ethnographie*. WZB Discussion Paper FS II, 98–112. Berlin: WZB.

Hirschmann, A. O. 1974. *Abwanderung und Widerspruch*. Tübingen: Mohr Verlag.

Jarren, O. 1998. "Medien, Mediensystem und politische Öffentlichkeit im Wandel." In *Politikvermittlung und Demokratie in der Mediengesellschaft*, ed. U. Sarcinelli, 74–96. Wiesbaden: Westdeutscher Verlag.

Jones, S. G. 1995. "Understanding Community in the Information Age." In *CyberSociety: Computer-Mediated Communication and Community*, ed. S. G. Jones, 10–35. Thousand Oaks: Sage.

Kleinsteuber, H. J. 1996. *Der "Information Superhighway": Amerikanische Visionen und Erfahrungen*. Opladen: Westdeutscher Verlag.

Leggewie, C., and C. Maar. 1998. Internet @ Politik: *Von der Zuschauer—zur Beteiligungsdemokratie*. Köln: Bollmann Verlag.

Luhmann, N. 1971. "Die öffentliche Meinung." In *Politische Planung*, 9–33. Opladen: Westdeutscher Verlag.

———. 1996. *Die Realität der Massenmedien*. Opladen: Westdeutscher Verlag.

———. 1997. *Die Gesellschaft der Gesellschaft*. Frankfurt am Main: Suhrkamp Verlag.

Maresch, R. 1997. "Öffentlichkeit im Netz." In *Mythos Internet*, eds. S. Münker and A. Roesler, 193–212. Frankfurt am Main: Suhrkamp Verlag.

Poster, M. 1995. *The Second Media Age*. Cambridge: Suy Press.

Rheingold, H. 1994. *Virtuelle Gesellschaft: soziale Beziehungen im Zeitalter des Computers*. Bonn: Addison Wesley.

Schmid, U. 1997. "Medien—Innovation—Demokratie: Zu den Entwicklungs— und Institutionalisierungsprozessen neuer Medien-Kulturen." In *Virtualisierung des Sozialen: Die Informationsgesellschaft zwischen Fragmentierung und Globalisierung*, eds. B. Becker and M. Paetau, 81–102. Frankfurt and New York: Campus Verlag.

Schmidt, S. J. 1994. "Die Wirklichkeit des Beobachters." In *Die Wirklichkeit der Medien: Eine Einführung in die Kommunikationswissenschaft*, eds. K. Merten, S. J. Schmidt, and S. Weischenberg, 3-19. Opladen: Westdeutscher Verlag.

Seligman, A. B. 1992. *The Idea of Civil Society*. New York: Free Press.

Thompson, J. 1995. *The Media and Modernity*. Cambridge: Polity Press.

Wehner, J. 1997. "Interaktive Medien—Ende der Massenkommunikation?" *Zeitschrift für Soziologie* 26: 96–114.

Wellman, B., J. Salaff, D. Dimitrova, L. Garton, M. Gulia, and C. Haythornthwaite. 1996. "Computer Networks as Social Networks: Collaborative Work, Telework, and Virtual Community." *Annual Review Sociological* 22: 213–38.

Wellman, B. 1997. "An Electronic Group is Virtually a Social Network." In *Culture of the Internet*, ed. S. Kiesler, 179–208. Mahwah, NJ: Lawrence Erlbaum Associates.

Wetzstein, T., et al. 1995. *Datenreisende: Die Kultur der Computernetze*. Opladen: Westdeutscher Verlag.

National Level Culture and Global Diffusion: The Case of the Internet

Carleen F. Maitland and Johannes M. Bauer

Introduction

The increasing globalization of the worldwide economy has led to increased emphasis on the international diffusion of technologies. One the most quickly diffusing technologies in the past decade are those involved with computing and communication. Interactive networks, such as the Internet and wireless telephony, in addition to computing hardware and applications, are being adopted at phenomenally rapid rates across the globe.

In order to identify the drivers of this rapid adoption a wide body of research has developed. The theoretical basis of this research and the range of variables used in these studies require clarification and extension, respectively. First, from a theoretical perspective a broad theory that permits a wide range of factors including economic and innovation-related features as well as cultural variables is needed. Typically, Diffusion of Innovation theory is used, however, with little discussion of the ramifications of using this originally individual-level theory for studies concerning global diffusion. Second, diffusion studies do not typically include cultural factors. There is thus a need to discuss the theoretical fit of cultural factors in global level diffusion studies as well as empirically testing their significance.

The research presented below examines these theoretical and empirical issues. Questions addressed include the following.

- Can Diffusion of Innovation theory be applied at the global level?
- What is the expected theoretical significance of culture on global diffusion?

- What are the results of an analysis of the impact of culture in global diffusion using the Internet as a case study?

This chapter is organized as follows. First, the theoretical questions concerning global diffusion and culture will be presented. Second, the findings of a case study concerning the role of culture in global Internet diffusion will be presented. Finally, the implications of the findings of the case study for culture in future global diffusion studies will be addressed.

Culture and Diffusion

In this section the role of culture in Diffusion of Innovation theory will be discussed. The section begins with a discussion of culture, its quantification, and levels of analysis. This is followed by a discussion of diffusion theory and its use in research at various levels of analysis.

Culture

Prior to embarking on a discussion of the issue of national culture, the term culture must first be defined. Culture is such a broad construct that the best one can do is place boundaries on its meaning for a particular application. Geertz (1973) defines culture as:

> . . . an historically transmitted pattern of meanings embodied in symbols, a system of inherited conceptions expressed in symbolic forms by means of which men communicate, perpetuate, and develop their knowledge about and attitudes toward life. (89)

More simply, culture can generally be described as the way of life of a people (Rosman and Rubel 1995). Specifically, it refers to the socially learned behaviors, beliefs, and values that the members of a group or society share. Certain cultural features, known as cultural universals, are present in each society. These universals include language and other symbols, norms and values, and the tension between ethnocentrism and cultural relativism. Ethnocentrism refers to the belief that one's culture is superior to all others, while cultural relativism requires that the value of customs and ideas of a society must be judged from within the context of that society (Persell 1984).

Although culture is defined as a societal-level construct, it certainly has implications for individual behavior. Culture can be seen as a mediator between human nature, which is universal, and personality, which is specific to the individual (Hofstede 1997). The result is that although a range of personality types will be found in any society, there will also be a preponderance of individuals with a particular kind of personality (Rosman and Rubel 1995). The personality type represents how people within a society respond to their cultural norms. This demonstrates the ability to draw conclusions about societal culture based on responses of a sample of individuals from a society.

National Cultural Characteristics

This study raises two somewhat controversial issues relating to the study of culture and technology. The first is whether or not culture can be quantified, and the second is whether one can speak of national-level cultural characteristics. Each will be addressed in turn.

The literature on culture is vast and here we will attempt to provide a rough sketch of the various perspectives. Some researchers view culture as an unmeasurable construct. This view may stem either from the perspective of the researcher vis-à-vis the group being studied or may be merely be a function of the lack of depth that quantification typically reflects. On this view, culture can be described but not quantified. First, in relation to perspective, if culture is embedded or reflected through cultural norms, then it is almost impossible to truly understand those norms from a position outside that particular "culture." However, once inside, the perspective changes and it is difficult to recognize what is different "culturally" about any group of individuals. Second, exacerbating these challenges of perspective, attempts to quantify culture are clumsy. Measures, through surveys for example, miss the subtlety of the cultural traits. The culture of a group may be seen as a combination of a variety of cultural traits. Cultural traits may have intricately interdependent relationships and attempts to measure individual traits in various societies (if indeed culture is a societal-level construct) obfuscate these interdependencies and insufficiently identify cultural differences.

A further complication in measuring culture exists in identifying the unit of analysis. Some see culture as a multi-level construct. We can talk of societal culture, organizational culture, or even family culture. In terms of accuracy of cultural descriptions, there seems to be a bias toward smaller entities. Generally, the smaller the size of the group the more accurate the description of its culture. This stems from the fact that it is simply less likely within a small group to identify a person who violates the cultural norms, thus potentially invalidating the description of that culture. Therefore, describing the cultural differences between two neighborhoods will be more accurate than describing the cultural differences between two societies. The problem lies with the "ecological fallacy" (or fallacy of division)—the impulse to apply group or societal level characteristics onto individuals within that group. The result for research on culture is that as we move up in levels of analysis the descriptions, either qualitative or quantitative, become increasingly difficult to defend.[1] For example, if culture is considered a societal-level construct it can be argued that societies do not respect national-level boundaries. Several distinct societies can exist within a nation, such as in northern and southern Italy, for example. It can also be argued that societies span national boundaries, such as a French cultural base in parts of Belgium. The result is quantified measures of national culture can be considered controversial, both for being quantitative and for their level of analysis.[2]

Despite these recognized limitations of quantified measure of national culture, this research will attempt to use these measures in a study of global diffusion. Although a society is more likely a better unit of analysis than a nation, there is very little quantitative data that can be found describing various cultural traits of societies around the globe. In addition to this, other data with which to compare the impact of culture is also rarely available on a societal level. National governments collect data that are often only relevant at the national level. Thus, studies of global diffusion are forced not from theory but from available data to use nations as the unit of analysis. The result is that studies using national level cultural characteristics must be careful in interpreting results. In particular, national level characteristics must not be interpreted at the individual level.

One might ask whether or not it is worthwhile to engage in research where the variables are potentially poor indicators of the theoretical construct involved. The reality is that all measures at the national level are suspect. Economic indicators such as Gross Domestic Product, employment, teledensity, infant mortality rates, etc., can all be called into question in terms of their accuracy of measurement (reliability) as well as how well they reflect what they are intending to measure (validity). Such is the nature of inter-national research.[3] With these caveats in mind, a discussion of potential national cultural dimensions follows.

Studies attempting to identify national cultural characteristics are plentiful. One of the most widely used sets of national cultural characteristics are those established by Geert Hofstede (1980). Hofstede analyzed survey data from an international sample of IBM employees from 1967 to 1973.[4] The survey questions were designed to measure work-related values. Hofstede used these measures of values, which are a component of culture, to identify national level cultural characteristics common among all of the respondents. He then created scales that provided a score on each of the characteristics for each of the fifty-one countries represented in the sample.

Hofstede found national cultures vary on five dimensions: individualism vs. collectivism, femininity vs. masculinity, long-term vs. short-term orientation in life, power distance, and uncertainty avoidance (Hofstede 1997).[5] For four of the five national cultural dimensions the implications for diffusion of interactive networks are inconclusive. Based on descriptions of the dimensions, contradictory hypotheses predicting both an increase and decrease in the speed of diffusion can be developed. This does not imply that the cultural dimensions are irrelevant for the study of interactive network diffusion. It does, however, highlight the need for a theoretical structure to more accurately predict the direction of the relationships.[6]

The one dimension that is theoretically unambiguous in terms of its implications for diffusion of interactive networks is uncertainty avoidance. The uncertainty avoidance dimension is reported as an index and is interpreted as the extent to which the members of a culture feel threatened by uncertain or unknown situations. In countries with low uncertainty avoidance (Jamaica, Denmark) it is common that motivation comes from achievement, esteem or belongingness; there is a high tolerance for deviant or innovative ideas and behavior. In strong uncertainty avoidance countries (Greece, Portugal) there is resistance to innovation and motivation for work comes from security as well as esteem and belongingness. The implications of uncertainty avoidance for diffusion of an innovation are clear. In low uncertainty avoidance cultures new ideas will be more readily accepted than in high uncertainty avoidance cultures. Thus, low uncertainty avoidance cultures should experience faster rates of diffusion of new technologies.

In addition to using generic national cultural variables, research by DeKimpe, Parker, and Sarvary (1997) suggests the variables used in a diffusion study should match the innovation being studied. In their study of the global diffusion of cellular telecommunications, they include a social system variable and social heterogeneity, measured as the number of ethnic groups in a country. Heterogeneity of ethnic groups relates to society-wide communication and hence the use of mobile telephones. This variable was shown to have a significant impact on a country's adoption timing.

In a similar manner, the case study presented in this research will use both cultural variables related to general innovative capacity (uncertainty avoidance) and those with specific implications for Internet diffusion (English language ability). Other national level cultural characteristics that may have implications for Internet diffusion include the communication patterns (near/distant), relative roles of work and family, perceptions of the role of technology in home life, and shopping behaviors. In general, measures of these innovation specific cultural characteristics will be more difficult to find. However, diffusion researchers should attempt to use both general and innovation specific cultural characteristics in their research.

Diffusion

Diffusion of Innovations theory as presented by Rogers (1995a) proposes rates of adoption can be explained by five categories of variables: (1) perceived attributes of the innovation; (2) type of innovation decision; (3) communication channels; (4) nature of the social system; and (5) extent of change agents' efforts. Within each category a wide range of variables exists, and their level of analysis varies from individual (such as all of the "perceived attributes" variables) to system level variables (such as those in the "nature of the social system" category). These variables can be used to compare diffusion of different innovations and to compare the rate of diffusion of an innovation among communities with different economic, demographic, and cultural characteristics. Both within and among the five general categories of explanatory factors, there is overlap. The categories are not mutually exclusive and the question of which category a particular variable belongs to is sometimes open to interpretation. Culture plays either an implicit or an explicit role in each of these five categories of variables.[7]

Traditional diffusion studies relied heavily on the notion of "perceived attributes" of an innovation. Although the potential adopter's perceptions of the innovation are important, objective characteristics also play a role (Lin and Zaltman 1973). Network-based and interactive innovations, in particular, possess certain objective characteristics that differentiate them from stand-alone innovations. These differences include network externalities and the "critical mass" effect. Recognition of these differences has led to changes in Diffusion of Innovation theory (Rogers 1986; Rogers 1995b).

Global Diffusion

As stated above, traditional diffusion research addresses diffusion mostly from an individual-as-adopter perspective with an emphasis on perceived attributes measured at the individual level. The theory has, however, been applied to groups as well. The analysis here uses diffusion theory and applies it to a global level analysis where nations are seen as adopters. The leap in levels of analysis requires attention to several questions. In what ways do nations display characteristics similar to individuals in areas that are relevant to diffusion? What are the mechanisms by which the individual characteristics affect diffusion and are these mechanisms applicable to a global level analysis? We will begin by addressing the most transferable aspects first, followed by a discussion of those characteristics or mechanisms which are not easily transferred.

Studies of innovation diffusion have identified characteristics of early adopters as having greater wealth, higher levels of education, and greater exposure to mass media (Rogers 1995a). At the individual level, persons with greater wealth have the financial resources to invest in new technologies, even before the advantages of the innovation are clear and well established by other adopters. Early adoption involves risk and those with greater financial resources are better able to afford these risks. In a comparative analysis of nations, wealth is measured by per-capita GDP. In a global level analysis the "adoption by a country" is usually really an adoption by an individual in that country. There are national-level institutions, however, that will affect the ability of an individual to make that adoption decision, by influencing their access to wealth, education, and mass media. Thus, national-level indicators of wealth, education and mass media should predict adoption just as individual-level measures would. The national measures are a mere aggregation of individuals' wealth, education, and mass media exposure. Nations with higher incomes, higher levels of education, and greater numbers of mass media channels can thus be reasoned to have higher levels of adoption through the same mechanisms as individuals.

That national level characteristics and national level adoption can be associated is the easier part of applying the individual level diffusion theory to a global analysis. The more challenging part of applying an individual level theory to a global level analysis comes from the more explanatory elements of diffusion theory.[8] These elements highlight the process of diffusion more so than the preconditions amenable to diffusion.

Above, the five categories of adoption variables were described as (1) perceived attributes, (2) type of innovation decision, (3) communication channels, (4) nature of the social system, and (5) extent of change agents' efforts. These categories of variables were derived from theories of interpersonal communication, mass media, sociology, etc., as well as evidence found in actual diffusion studies. The first category "perceived attributes of the innovation" emphasizes that the perceptions of individuals are what drive diffusion, as opposed to some externally defined objective characteristics. Perceived attributes have been identified as being in one of five categories: relative advantage, compatibility, complexity, trialability, and observability. Although the emphasis has been on individuals' subjective perceptions of these traits, the categories of the traits themselves, such as trialability and observability, suggest there is some role for objective traits in diffusion studies.

The role of these objective traits may be conceptualized as follows. Perhaps the attributes of an innovation related to its adoption can be broken down into two components: objective attributes and subjectively perceived attributes. The adoption of various innovations will rely on a mix of these two parts, and the weights assigned to each part will vary with the innovation. If one accepts this objective/subjective innovation attribute argument, it is possible to move the theory from one constrained to the individual (subjective) mode to one where higher level units of analysis can be considered. If there are objective traits that make a technology more advantageous in some environments than others then it is possible to apply diffusion theory to the study of organizations, groups, firms, and nations.

This role for objective innovation attributes is further supported by the above discussion of the need to modify diffusion theory to take into account the objective traits of interactive networks. For an interactive network-based innovation, the objective characteristics (the value of the innovation relying on the number of other adopters, the need for large-scale investment for network technologies, etc.)

play a role in adoption. Thus, although not explicitly stated by Rogers, the need to "adjust" diffusion theory for interactive networks points to a need for a two-fold conceptualization of innovation attributes for diffusion theory.

The usefulness of this dual conceptualization is further demonstrated by the following hypothetical example, which is concerned with the specific innovation attribute of compatibility. In a theoretically derived study of the adoption of NetMeeting software it would be possible to explain, without considering subjective perceptions of compatibility, why workgroups who use Microsoft Explorer are more likely to adopt NetMeeting than those who use Netscape. The compatibility of NetMeeting with Microsoft Explorer is an objective (very well-planned) trait. By limiting the study to only objective traits the study admittedly misses information, namely the impact of perceived attributes. However, by concentrating only on objective characteristics it may be possible to take a broader perspective on the issue. Thus, by "going up" levels of analysis, away from the individual, overall adoption information is sacrificed. However, the information derived from macro-level studies may help inform a different audience as well as lay the groundwork for more focused individual level research.

The preceding discussion has laid the basis for our study of culture in global diffusion. The nature of culture was considered along with issues of levels of analysis and measurement in the use of cultural variables. The use of culture in global diffusion studies was also discussed. This discussion led to a call for a reconceptualization of attributes within diffusion—most importantly, our emphasis on objective traits and attributes—that would then allow diffusion theory to be applied at higher levels of analysis. Thus this section concludes with a theoretical basis for global diffusion studies that includes culture as a variable driving this diffusion. In the following section a case study of Internet diffusion will demonstrate how diffusion theory, as revised in these ways, can be applied at the global level and how quantitative measures of national culture can be included in such an analysis.

The Case of Global Internet Diffusion

The following case demonstrates how diffusion theory can be applied at the global level. The study of Internet diffusion attempts to identify the driving factors in this diffusion across the globe. With nations as the unit of analysis, a broad range of variables, including culture, are identified as possible explanatory variables. In the following sections the variables and sources of data are first discussed. Subsequently, the types of analysis and results are presented. Finally, we discuss the implications of the results for future studies involving culture.

Research Questions, Variables, and Data

This research proposes to examine Internet diffusion at the global level. Using the theoretical concepts present above two general research questions arise:

- Which diffusion variables explain differences in adoption between countries? Do these same variables explain diffusion within countries?
- How do cultural variables compare to economic and infrastructure variables in terms of their explanatory power?

Previous studies of Internet diffusion and our theoretical discussion above suggest that a wide range of competing and interdependent forces are driving the complex Internet diffusion process. Although the role of culture in this process is interesting on its own, our question about culture is, how does it compare to other more commonly used attributes for describing ICT diffusion such as wealth and technical capabilities? In this global level analysis, explanatory variables will be represented by national level indicators, including national level measures of culture. To simplify the analysis the explanatory variables will be grouped into categories representing infrastructure, the national economy, and national culture. Their roots in diffusion theory are mapped onto specific categories of variables and are shown in Table 1. A more in-depth discussion of the variables follows.

VARIABLES

Diffusion studies have shown that relative advantage, usually in terms of cost, and compatibility is the most powerful for predicting adoption (Tornatzky and Klein 1982; Eastlick 1993). It will not be surprising, therefore, if in this study high GDP, which makes the

Table 1
Predictors and Their Roots

Category	Variable	Theoretical Root
Economy	GDP per capita	Relative advantage
	Access to mass media	Communication channels
	Education	Nature of the social system
	PC's per capita	Compatibility
	Trade	Communication channels
Infrastructure	Centrality	Communication channels
	Teledensity	Compatibility
	Cost of access	Relative advantage
	Peripherals	Compatibility
Culture	Uncertainty avoidance	Nature of the social system
	Gender equality	Nature of the social system
	English language	Nature of the social system

cost of Internet adoption relatively low and hence creates a higher relative advantage, is a strong predictor. In addition to GDP, the level of education and a nation's exposure to mass media will also facilitate the adoption of the Internet. All three of these variables are characteristics of early adopters of communication technologies in general (Rogers 1986). Education provides the necessary skills while the mass media perform the function of spreading information about the innovation. Another mechanism for the mass media to influence Internet adoption is by enhancing its prestige. For example, as increasing numbers of firms place their URLs in advertisements, both in print and on television, this puts pressure on other firms to "keep up."

One of the most intuitive variables for explaining Internet diffusion is the presence of personal computers. The existence of PCs and familiarity with their operation will enhance the compatibility of the Internet through previous experience. Despite this variable's intuitive appeal, previous research has shown its explanatory power to be less than expected. In countries belonging to the group the Organization for Economic Cooperation and Development (OECD), it is true that high Internet penetration is accompanied by high computer penetration; however, there are some countries (Switzerland for example) where high computer penetration has not been translated into Internet connections (OECD 1996).[9]

The final variable in the economic category will represent the role of an individual nation in the social structure of the global system. It is expected, based on the role of communication networks in the diffusion process, that a country more highly integrated in the global economy through trade will more quickly adopt this innovation.

The second broad category of variables is infrastructure. Variables in the infrastructure category reflect the nature of the telecommunications infrastructure but also represent the communication structure between countries. The telecommunications infrastructure of a nation facilitates international communication which in turn drives diffusion. The centrality of a nation in the global communication network is expected to influence its adoption time as the advantages of the innovation will be spread through international communication channels (Allen 1988; Rogers 1995; Valente 1995). The second variable expected to explain Internet diffusion is teledensity, a measure of the number of telephone lines per capita. As previously mentioned, access to a local telecommunication network, particularly in developing countries, as essential to the continued growth of the Internet (Blumenthal 1997). Access is however a necessary but not a sufficient condition for Internet diffusion. The price of access will also play a role in determining the relative advantage of the medium. Paltridge (1996) showed that a strong relationship exists between competitive telecommunication markets, price of both local and leased access, and Internet diffusion for the OECD countries. He states, "on average, the penetration of Internet hosts is five times higher in competitive than monopoly markets" (Paltridge 1996, 26). The lack of available bandwidth and high cost of leased lines have been cited as one factor responsible for the slow expansion of the Internet by a Japanese ISP (Hahne 1997).

The final variable in the infrastructure section represents a country's orientation toward using network peripherals, such as modems or fax machines. This reflects use of the telecommunications network as a data network. These innovations create greater compatibility through previous experience.

As stated earlier, this model consists of three categories, the last of which is culture. Culture is seldom considered in quantitative analyses of diffusion at the global level, although some notable exceptions have appeared in the area of marketing (see Parker 1994 for a review). Results of one study, that of DeKimpe, Parker, and Sarvary (1997), suggest the cultural variables should match the innovation being studied. Thus, in this study both general cultural variables and those specifically related to the Internet will be used.

The first variable used to reflect cultural influence on the adoption timing of a country is uncertainty avoidance. The implications of uncertainty avoidance for diffusion of an innovation are clear. In low uncertainty avoidance cultures new ideas will be more readily accepted than in high uncertainty avoidance cultures. Thus, low uncertainty avoidance cultures should experience faster rates of diffusion of new technologies.

The second cultural characteristic expected to affect diffusion of the Internet is gender equality. In his study of the sourcing of innovations Herbig (1994) suggests gender equality will impact a country's innovativeness. His rationale for use of this cultural trait in explaining the source of innovativeness is simple. A country in which gender equality is low fails to tap the potential of half its population, thus reducing its potential for innovation. In high gender equality countries, the potential for innovation is greater because a larger percent of the total population are in positions to innovate. The use of this variable is similar to the sex roles variable used by Gatignon et al. (1989).

The last variable in the culture category is that of English language ability. Although the previous two cultural variables are related to diffusion of technical innovations in general, English language ability is a cultural variable related specifically to Internet adoption. It has been observed that while languages such as Spanish and Japanese are gaining popularity on the Internet (Marriot 1998), the ability to speak English will certainly impact the relative advantage the medium presents. The orientation of a nation toward English is also important to Internet diffusion because some of its earliest adopters are English-speaking and therefore nonadopters who speak English are more likely to be influenced by early adopters.

In the above section the theoretical variables to explain Internet diffusion were presented. The list is certainly not exhaustive and has been constrained by data availability. The data that are available and serve as measures of these variables are described below.

Data

The ability of international researchers to practice their craft is highly dependent on the availability of data. It is a challenge to collect data for a multivariate analysis on a global scale. The following section describes the sources of the data used in this study and, where possible, addresses issues concerning compatibility of data sources.

Dependent Variable

The growth of the Internet can be measured in a variety of ways including changes in the amount of network traffic it generates, the number of hits on web sites, estimates of the number of users, and the number of computers it connects. Methods of measuring its growth and impact are constantly being improved, with companies racing to establish their measurement system as the most reliable and accurate. The global nature of this study creates a unique requirement for measuring Internet growth. The measure of growth must be available for a wide range of countries including those in the developing world. The most widely reported statistic on growth is that of host counts.[10]

Although Internet host counts do not give accurate statistics on the number of users or their demographics, they do provide certain advantages. Data on the number of Internet hosts have been collected since the early eighties, and this consistency of measurement is one the method's main benefits (Rickard 1995). Counting the number of Internet hosts, which represent access to the Internet as opposed to usage or number of users, provides an estimate of the minimum number of Internet users. To create estimates beyond a minimum number a factor of users-per-host is required. This ratio is approximately seven in the OECD countries and is expected to decline, assuming extensive diffusion over several years, to one. By comparing estimates of the number of users gathered through random telephone survey studies with the minimum number of users from host counts, estimates of Internet access have improved, although no truly accurate measure exists (Lotter 1998).

Therefore, in the following analysis an "adoption" is considered the addition of a host to the Internet.[11] The data on the number of Internet hosts both globally and on a country-by-country basis were obtained from the Network Wizards web site (<http://www.nw.com/>), the most frequently cited source of Internet diffusion data (Press 1997). Although the Internet had its start back in 1969 as ARPANET and evolved into Bitnet in 1981 (Leiner et al. 1998), the data used in this study begin in 1991 and are measured in six-month increments. The reason for starting in 1991 is although data on the number of Internet hosts prior to 1991 exist, it was not until 1991 that hosts were attributed to a particular top level domain name representing their country of origin. Attributing hosts to particular domain names allows for a rough estimate of the number of hosts in each country. The data is further limited by a change in data collection techniques. After July 1997 the data collection techniques of Network Wizards changed, thus making comparisons with data collected previously questionable. Therefore, the dependent variable data used in this study start in July 1991 and are reported every six months until July 1997. The host counts indicate that as of July 1997, 185 of 239 total countries were connected to the Internet. With this data two sets of dependent variables are constructed. The first indicates when a country first adopted the Internet, and the second is concerned with the rate of growth of the Internet in each country.

The implications for using this data set in this analysis are the following. First, all countries adopting during and before July 1991 are lumped into this time slot. Compared with subsequent six-month periods, the non-cumulative number of adopters in this initial time period is distorted. This distortion is taken into account in both the analysis and conclusions made from the analysis. The second implication is the data set is truncated and the limited number of data points will have implications for the type of analyses possible. Despite these limitations, thirteen time points are certainly sufficient to draw some conclusions. The data on the aggregate global diffusion of Internet hosts is shown in Figure 1.

Figure 1
Number of Internet Hosts Worldwide

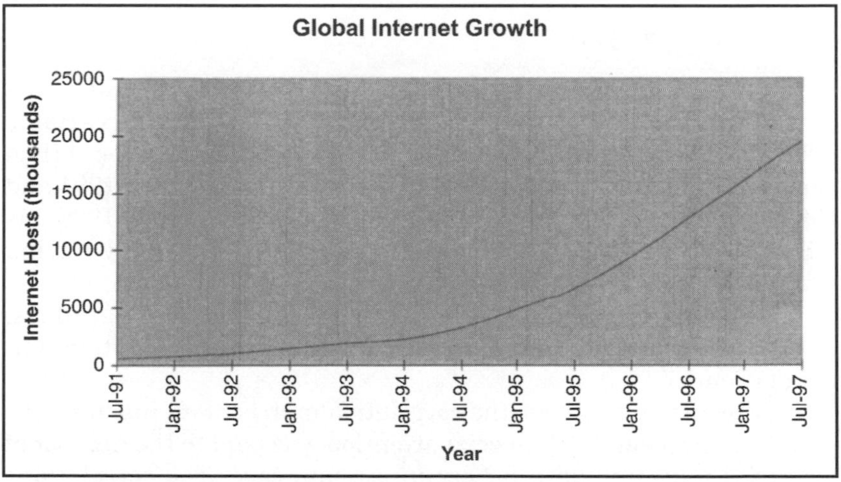

Source: Network Wizards <http://www.nw.com>)

Independent Variables

To explain and predict the diffusion of the Internet the independent variables suggested above will be used. Each of these variables is represented by one or more measures. It is nearly impossible to find national-level indicators for all countries in the world. The result is that one measure may be available for 85 countries while another is available for 110. It may also be the case that those 85 are not all included in the 110 of the other variable. In general, the year of the measure was chosen based on the number of countries reporting the data in the 1990–1995 time period. To maximize the number of variables available for the analysis it was necessary to take a cross-sectional measure for each variable and these cross-sectional data were taken from different years. For example, the income measure is from 1994 while the newspaper variable is from 1992. Although not ideal, the impact on the analysis should be minimal as the measure reflects the rank of country vis-à-vis the other countries and these rankings, in national level macroeconomic and demographic variables over a short time period, are fairly stable.[12] Also, use of multiple measures for one variable does not imply that scales will be formed. It merely adds flexibility in choosing a set of independent variables that are (1) not highly correlated and (2) represent the largest number of countries possible.

The variables were gathered from a variety of sources. The source, year of the data, number of observations, and expected correlation with the dependent variable are presented in Table 2.

Multivariate Analyses

In the first part of this analysis the independent variables discussed above will be regressed onto the dependent variable (START). The start variable was constructed by the number of time periods a country has had an Internet connection between 1991 and 1997. For example, a country having adopted in January 1996 has a START value of four, compared to twelve for a country having adopted in January 1992. The data are available for 185 countries. Thus, this is an analysis of the impact of independent variables, which are assumed to be stable over this time, on the adoption timing of countries.

To begin the analysis, the correlation matrix of all independent variables was examined. Special attention was paid to the number of countries reporting data for each bivariate case. The correlations demonstrate the high multicollinearity between variables. For example, GDP, an economic variable, is highly correlated with two infrastructure variables, the number of fax machines (.914, $p<.01$) and teledensity (.885, $p<.01$). The gender empowerment measure is highly correlated with teledensity (.689, $p<.01$) and PCs per thousand (.692, $p<.01$). Thus, the empirical analysis will be performed under the limitation of multicollinearity and the number of countries reporting data.

Before continuing with the analysis it is interesting to look at the explanatory power of a few variables from each category to see if one group dominates the others. From Table 3 it can be seen that the cultural variables are slightly less powerful predictors of start time than the economic and infrastructure variables. This was expected conceptually and supports previous diffusion research that has found economic factors to be the strongest predictors of adoption. The position of this research is not to show that cultural variables are the strongest predictors but that they can help in explaining adoption and this position is supported by this first step in the analysis.[13]

Due to the constraints of multicollinearity and data availability, two methods for creating viable models (or combinations of IVs) were used. The first method identifies a dominant independent variable, with subsequent independent variables (IVs) being added based on their relationship with the dominant IV. This first method sacrifices comparability between models for including a wider range of variables and potentially a larger number of countries in the analysis.

The results of this method are models that explain a comparatively larger amount of variance in the dependent variable, however the models are not directly comparable with one another because each combination of independent variables creates a unique sample of countries reporting data for those particular variables. This problem is addressed by the second method. The second method uses a sub-sample of fifty-five countries, all of which report data for eight of the strongest predictors. This method allows comparisons across models to be made.

Inter-Country Differences

The results of methods 1 and 2 are given below; the correlation matrices for both the full sample and the sub-sample of countries, as well as a list of the countries in the sub-sample, are provided in

Table 2
Independent Variables

Independent Variables	Measures	Year	Description	N	Expect *r*
Economy					
Per Capita GDP	GDP_CAP	1994	Per capita Gross Domestic Product; Source: ITU World Telecommunication Indicators (WTI) Database (1997)	153	+
Trade	TRADE	1994	Trade as a percent of GDP; Source: World Bank (1997)	147	+
Research Net	DEFENSE	1992	Defense Expenditure as % of GDP; Source: World Bank (1997)	75	+
Access to Mass Media	NWSPAPR	1992	Daily Newspapers per 1,000; Source: World Bank (1997)	154	+
Education	SCHLNROL	1990	School Enrollment, tertiary (% of gross) Source: World Bank	110	+
	EDUCBUD	1990	Public Spending on Education, tertiary (% of GDP) Source: World Bank (1997)	75	+
Personal Computers	PCPTHO95	1995	PCs per thousand; Source: ITU WTI Database (1997)	87	+
Infrastructure					
Teledensity	TELEDEN	1994	Mainlines per 1000; Source: ITU WTI Database (1997)	200	+
Cost of Access	CLCOST94	1994	Cost of a 3-minute local call; Source ITU WTI Database (1997)	128	–
	LESDPC94	1994	Leased lines per capita; Source: ITU WTI Database (1997)	99	+
	COMPET	1997	Market structure: monopoly or competition; Source: Mody, Maitland et al. (1997)	84	+

Centrality	INTCALMN	1994	International telecom, outgoing traffic (minutes per subscriber); Source: World Bank (1997)	170	+
	INTCALL	1995	International telecom, average price call to USA (US$ per 3 min.); Source: World Bank (1997)	139	–
	CENTRAL	1989	Centrality closeness measure using NEGOPY; Source: Sun and Barnett (1994)	93	–
	LINKS	1989	Number of 'nominations'; Source: Sun and Barnett (1994)	93	+
Peripherals	FAX1993	1993	Estimated number of fax machines per 1000 people; Source: World Bank (1997)	105	+
Culture					
	GEMPWR2	1996	Reverse coded 'Gender Empowerment' Score; Source: UNDP (1996)	92	+
	UAI	1980	Uncertainty Avoidance Score; Source: Hofstede (1997)	50	–
	TOEFL	1996	Test of English as a Foreign Language; Source: Educational Testing Service, New Jersey, USA (1996)	167	+

Appendix A. For both analyses the explanatory power of the independent variables was tested using stepwise regression with the most highly correlated variable with the dependent variable being entered first. Variables added to the model were entered only if their intercorrelation was 0.6 or less. Using these selection criteria for additional variables, stepwise regression then makes clear the amount of variance attributed to each new variable in the model.

Method 1

For the full sample the most highly correlated variables with the dependent variable START in order are Newspapers per one hundred, GDP per capita, Teledensity, Gender Empowerment, International Call Cost, School Enrollment, PCs per one thousand, English Language Ability, Links, and Centrality. Exploring the correlation matrix, seven unique models were found.

The model with the strongest explanatory power includes the variables Teledensity, International Call Cost, and English Language Ability (TOEFL). In addition to having the largest R^2 (.614)[14], the model was tested on the largest number of countries (122). The results, particularly regarding the strong explanatory power of teledensity, were expected. Unfortunately, it was impossible to combine teledensity with other variables due to the high intercorrelations. GDP per capita also suffered the same fate. The result of these high intercorrelations is that the variables' power can be compared with only a few other variables.

Table 3
Explanatory Power of Individual Categories

Variables	Adjusted R^2	R^2 Change	Betas (sig. $p < .01$)	N
Economic	.476			97
GDP_CAP		.390	.392	
SCHNROL		.097	.389	
Infrastructure	.434			74
CENTRALITY		.402	−.569	
INTCALL		.048	−.228	
Culture	**			
GEMPWR	.341	.349	.590	92
TOEFL	.277	.281	.530	167

**The simultaneous inclusion of both variables was not possible so each variable was regressed individually.

Examining the relative power of variables within models by the standardized betas the following results are obtained (table 4).[15] In models 1a, 1b, 2b, 5, and 6 it can be seen that International Call Cost is a stronger predictor than Newpapers, English Language Ability (TOEFL), GDP per Capita, PCs per thousand, and Links. Models 3 and 4 respectively show that Teledensity and Gender Empowerment are stronger predictors than International Call Cost. A direct comparison of Teledensity and Gender Empowerment was not possible due to multicollinearity (r=.689). However, the partial correlations of each variable with the dependent variable controlling for one another would indicate Teledensity has a stronger relationship with START than does Gender Empowerment.[16] Model 2a is of interest because it contradicts the other models showing School Enrollment to have greater explanatory power than International Call Cost. Model 2b shows the English Language Ability factor to be almost as powerful as that of GDP, with respective standardized betas of .185 and .201. Therefore, the top three overall predictors are Teledensity (ΔR^2=.526), Gender Empowerment (ΔR^2=.468), and International Call Cost (ΔR^2=.438).

Method 2

The results obtained from the second method are as follows (table 5). In method 2 a sub-sample of fifty-five countries was used. The sub-sample includes both developed and developing countries, those with old and recent adoptions, and represent a variety of cultural backgrounds. For the reduced set of countries, Teledensity, International Call Cost, and School Enrollment are the top three most highly correlated variables with the dependent variable, START. These variables were combined with other variables based on their own correlation with the dependent variable and their relationship to one another. This process produced the following models and results.

With this reduced sample the model with the highest explanatory power (R^2=.546) includes International Call Cost and School Enrollment as the predictors. The strongest model in the first section includes Teledensity, International Call Cost, and English Language Ability (TOEFL). Unfortunately, in this smaller sample of countries the Teledensity and International Call Cost variables are too highly correlated (r>.6) to test this as a model. However, as with the previous analysis, when Teledensity was combined with TOEFL, the only variable able to be paired with Teledensity, Teledensity was the stronger predictor. Also, International Call Cost is

Table 4
Full Sample Models for START

Model	N	Adjusted R^2	ΔR^2	β	ρ
Model 1a	84	.553			
NWSPAPR			.392	.279	<.01
INTCALL			.123	−.344	<.01
SCHLNROL			.054	.308	<.01
Model 1b	80	.575			
NWSPAPR			.394	.235	<.05
INTCALL			.138	−.331	<.01
SCHLNROL			.044	.265	<.01
TOEFL			.021	.170	<.1
Model 2a	88	.552			
SCHLNROL			.387	.341	<.01
INTCALL			.145	−.339	<.01
GDP_CAP			.035	.244	<.05
Model 2b	83	.578			
INTCALL			.399	−.329	<.01
SCHLNROL			.141	.287	<.01
TOEFL			.035	.185	<.05
GDP_CAP			.023	.201	<.05
Model 3	122	.614			
TELEDEN			.526	.502	<.01
INTCALL			.068	−.255	<.01
TOEFL			.030	.199	<.01
Model 4	78	.548			
GEMPWR			.468	.495	<.01
INTCALL			.092	−.358	<.01
Model 5	79	.467			
INTCALL			.438	−.509	<.01
PCPTHO95			.042	.257	<.05
Model 6	64	.506			
INTCALL			.388	−.398	<.01
NWSPAPR			.116	.369	<.01
LINKS			.026	.170	<.1
Model 7	76	.361			
NWSPAPR			.304	.492	<.01
CENTRAL			.074	−.278	<.01

again a stronger predictor than School Enrollment, and School Enrollment is stronger than GDP per capita. The reduction in the sample size resulted in the bivariate correlation of Newpapers with START going from .637, the highest correlation in the first analysis, to .565, only the fifth strongest variable. As in the previous analysis it has less explanatory power than International Call Cost, and

Table 5
Reduced Sample Models for START

Model	N	Adjusted R^2	ΔR^2	β	ρ
Model 1	55	.454			
TELEDEN			.444	.580	<.01
TOEFL			.030	.194	<.1
Model 2	55	.546			
INTCALL			.444	−.512	<.01
SCHLNROL			.119	.378	<.01
Model 3	55	.435			
GDP_CAP			.345	.280	<.01
SCHLNROL			.086	.358	<.05
TOEFL			.035	.210	<.1
Model 4	55	.516			
INTCALL			.444	−.517	<.01
NWSPAPR			.090	.335	<.01
Model 5	55	.364			
SCHLNROL			.345	.469	<.01
FAX1993			.043	.238	<.1
Model 6	55	.356			
GDP_CAP			.329	.464	<.01
TOEFL			.051	.251	<.05
Model 7	55	.394			
SCHLNROL			.345	.388	<.01
NWSPAPR			.072	.334	<.05

model 7 shows that in a direct comparison with School Enrollment, it has less explanatory power.

Growth in Individual Countries

The final part of this analysis is concerned with the factors influencing growth in individual nations. International diffusion studies can be concerned with breadth, the diffusion across countries, or in its depth, the diffusion of the innovation within each country. When examining depth the dependent variable will be a measure of growth.

Not all countries have had the Internet for a sufficient length of time to judge their growth. By first taking those countries with ten or more data points and removing outliers, a sample of forty-eight countries was created. These countries, because they are earlier adopters of the Internet comparatively, do not constitute a representative sample of the countries used in the analysis above. In fact, they are expected to score higher on the variables found to be significant predictors of adoption timing (lower international call costs, higher teledensity, higher school enrollment). By using this sub-sample it may be possible to control for the influences of these variables.

Of these forty-eight countries some adopted the Internet earlier than others. To reflect this aspect of growth a measure was devised that took the percent adoption achieved by 1997 and multiplied it by the reverse-coded START variable. Therefore, countries achieving higher adoption percentages in fewer number of years will have a higher growth score. The growth measure also controls for country size as the percentage adoption figure is based on the number of phone lines which partially reflects a country's size.[17]

The growth measure is most easily understood by examining a few diffusion curves and their scores. Figure 2 displays diffusion curves of six countries, four having started their growth in 1991 (France, Germany, the UK and US), the Philippines in 1992, and Costa Rica in 1993. The graph shows that of the four countries that started their growth in 1991 (France, Germany, the UK and US), the growth figures are highest for the country having achieved the highest adoption percentage (the US). Comparing the growth measures for France and the Philippines one finds they are both 0.02, although on the graph the growth of France is a bit higher. They are given the same growth score because France, although having achieved higher adoption, had a longer period of time to do so. On the graph France and Costa Rica have nearly the same diffusion pattern. However,

Figure 2
Comparing the Growth Variable Among Countries

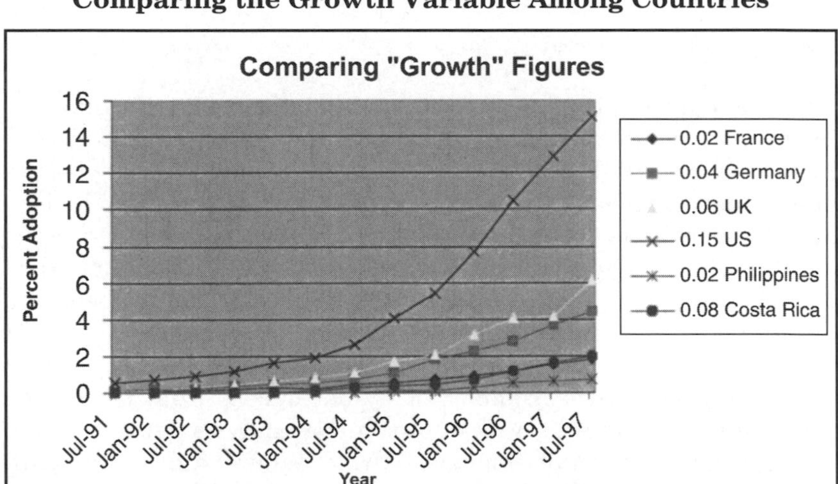

Table 6
Full Sample Model Testing for GROWTH

Model	N	Adjusted R^2	ΔR^2	β	ρ
Model 1	31	.395			
PCPTHO95			.355	.421	<.05
EDUCBUD			.081	.334	<.1
Model 2	33	.462			
GEMPWR			.319	.441	<.01
UAI			.177	−.438	<.01
Model 3	26	.441			
UAI			.413	.522	<.01
EDUCBUD			.072	−.295	<.1
Model 4	30	.356			
EDUCBUD			.303	.491	<.01
NEWSPAPR			.098	.318	<.05
Model 5	37	.484			
PCPTHO95			.389	.473	<.01
UAI			.124	−.383	<.01

Costa Rica started two years after France and is thus given a score higher than that of France (0.08 versus 0.02).

Once again the trade-off between including a larger number of countries in the analysis versus comparability between the models must be made. Therefore, the following analysis will proceed as in the previous multivariate analysis where first models are tested using the entire sample of forty-eight countries and then models are further tested with the reduced sample. The forty-eight countries and the countries in the reduced sample are listed in Appendix A. According to method 1, the models in this first part of the analysis were formed based on empirical considerations. Also, it should be noted that the correlation between START and GROWTH was quite low ($r=.155$, n.s.), indicating that they are measures of different phenomena.

The most highly correlated variables with the dependent variable were PCs (.591), Gender Empowerment (.569), and Uncertainty Avoidance (−.556), all significant at $p<.01$. The bivariate relationships allowed five models to be tested (table 6). Once again the model with the greatest explanatory power was tested on the largest number of countries. PCs and Uncertainty Avoidance produced an adjusted R^2 of .484, with PCs being the stronger of the two predictors. PCs was also shown in model 1 to be a stronger predictor than Education Budget. Models 3 and 4 show that Uncertainty Avoidance is stronger than Education Budget but that Education

Budget is stronger than Newspapers. Although Uncertainty Avoidance is weaker than PCs, model 2 shows it is stronger than Gender Empowerment.

For the second part of the analysis, listwise deletion of variables with fewer than thirty-five observations resulted in a sample of twenty-eight countries (table 7). The most highly correlated variables with GROWTH were PCs (r=.614), Gender Empowerment (r=.562), and Uncertainty Avoidance (r=−.561), all significant at $p<.01$. Five possible models were tested and the one with the strongest explanatory power included PCs and Uncertainty Avoidance, once again with PCs being the strongest predictor. In model 2 Uncertainty Avoidance was again found to be the weaker predictor when compared with Gender Empowerment. A direct comparison of Gender Empowerment and PCs was not possible through regression analysis due to multicollinearity: but the partial correlations with GROWTH controlling for one another indicate PCs has greater explanatory power.[18] The remaining models show Uncertainty Avoidance to be a stronger predictor than Fax Machines, Teledensity, and GDP_CAP.

Results

From the first part of the analysis it was shown that, using the full sample of countries, cultural variables were less powerful for ex-

Table 7
Reduced Sample Model Testing for GROWTH

Model	N	Adjusted R^2	ΔR^2	β	ρ
Model 1	28	.461			
PCPTHO95			.377	.467	<.01
UAI			.124	−.381	<.05
Model 2	28	.455			
GEMPWR			.315	.442	<.01
UAI			.180	−.441	<.01
Model 3	28	.383			
UAI			.315	−.439	<.05
FAX1993			.114	.359	<.05
Model 4	28	.438			
UAI			.315	−.489	<.01
TELEDEN			.165	.413	<.01
Model 5	28	.388			
UAI			.315	−.491	<.01
GDP_CAP			.118	.351	<.05

plaining adoption timing than were typical economic or infrastructure variables. Bivariate relationships of the predictors and START found Newspapers, GDP, Teledensity, Gender Empowerment, and International Call Cost to have the highest correlations. These predictors represent all three economic, infrastructure, and cultural categories. When put into models based on empirical relationships the most powerful model included Teledensity, International Call Cost, and English Language Ability as predictors of adoption timing. The strong predictive nature of Teledensity was expected, however the role of International Call Cost is less intuitive. International Call Cost was a measure of network centrality where countries with higher costs for a three minute call to the US were considered less central to the global communication network. Equating network centrality with the cost of a call to the US is appropriate for a study of this innovation as the Internet originated in the US. Although the predictive strength of this variable was not expected, its negative relationship with the dependent variable was.[19] It should also be noted that the high correlations of Newspapers and GDP with the dependent variable did not translate into strong predictive variables. In models with other variables these variables were easily dominated by International Call Cost. The weaker power of the GDP variable was surprising given findings from previous research and intuition.

The prominent role of International Call Cost was also seen in the reduced set of variables. In this analysis, the highest bivariate correlations between the predictors and START, based on the reduced sample, were between START and Teledensity, International Call Cost, and School Enrollment. International Call Cost and School Enrollment together formed the most powerful model with R^2=.546. The strong role of education in the diffusion process is also of interest as it provides further support for the significant role of education in many forms of development.

The next part of the analysis examined the relationship between predictor variables and the growth of the Internet within countries. Although this process attempts to find the best model to explain growth within countries, this is not an implicit assumption that the diffusion processes in all countries are the same. It is recognized that the drivers of intra-country access growth will differ among countries. The models merely represent the most common driving forces across countries, having a better fit in some countries than others. The sample used for this analysis, countries having the Internet for at least five years, was naturally higher in those variables that explain adoption. The bivariate relationships between the independent variables and the dependent variable GROWTH showed a heightened impact of the cultural variables. The three most highly correlated variables were PCs, Gender Empowerment and Uncertainty Avoidance. In both the full sample and reduced sample analyses, the model with the strongest explanatory power included PCs and Uncertainty Avoidance. For the full sample R^2=.484 and for the reduced sample R^2=.461.

Conclusions

This research is concerned with the use of national level cultural variables in global diffusion studies. At the outset two issues for this type of research were raised. The first was concerned with the level of analysis and use of quantitative measures when studying culture. The second was concerned with applying diffusion theory at the global level. The case of global Internet diffusion demonstrated the use of quantitative national culture variables in a global diffusion study. The case also demonstrated how national level indicators can be used as proxies for traditional diffusion variables. Below a detailed discussion of the implications of the case for culture and global diffusion studies is presented.

The case of Internet diffusion was used as an example of global diffusion research. Overall, the case demonstrates the need for a consistent set of global-level indicators in all categories. This is no surprise to international agencies such as the World Bank who allocate significant resources to data collection efforts. It is to be hoped that in the future more consistent reporting of data will occur and will begin to include cultural variables. Perhaps someday as much effort will be devoted to measures of national culture as are devoted to measures of national economic growth. Maybe someday we will be able to look back on the present as an era of extremely primitive measures.

The case of global Internet diffusion also demonstrates how objective measures of innovation attributes such as compatibility can be used in global level analyses. In the case study, objective measures of innovation compatibility were reflected by proxies such as the use of personal computers per capita, teledensity, and the use of telecom network peripherals, such as fax machines. It was observed that when using these objective traits, part of the diffusion process—that part concerned with an individual's perception of the innovation—is not taken into account. It would be interesting to find through further research whether or not perceptions of compatibility of the Internet are influenced by use of personal computers or telecom network peripherals and the relative strength of both. If the results matched those found here, use of personal computers would create a greater sense that the Internet is a compatible innovation. Further research on the implications of findings of global level diffusion research for more micro-level diffusion research is still needed.

In terms of cultural factors, the case of global Internet diffusion also highlighted the need to use cultural variables that are specific to the innovation being studied. Although cultural variables measured across a wide range of countries are difficult to find, creativity can help a researcher identify appropriate sources. As expected, in comparisons with economic and infrastructure variables, cultural factors were less powerful in their explanatory significance. They did, however, increase the predictive power of models. In predicting the adoption timing of countries the cultural trait of English language ability was a factor in the most powerful model. In terms of intra-country growth both uncertainty avoidance and gender empowerment were important factors. Thus, English language ability appears to play a greater role in determining when countries first adopt the Internet, while uncertainty avoidance and gender empowerment play a greater role in intra-country growth. Although the general notion of these cultural measures is understood, precisely how they impact the process of Internet diffusion will require further research.

Overall, the case of global Internet diffusion demonstrates the feasibility of including quantitative measures of national cultural variables into a multivariate global study. The limitations of such research in terms of the level of analysis of culture and in its quantification are recognized. Despite these limitations, global diffusion research that includes cultural variables can serve as a first step in identifying trends in global diffusion and the role culture plays in this process.

Notes

1. It can be argued that this is not a problem unique to the measurement of culture. Culture does, however, face problems in that it is indivisible, like variables such as wealth or education.

2. This description is the result of debates on this topic during the conference of Cultural Attitudes Towards Communication and Technology (CATaC '98), London, 1998.

3. Furthermore, as noted by Dekimpe et al.: "A practical problem in testing 'global theories' is the need to use globally represented proxies. As applied international researchers are well aware, the requirement to use covariates which measure international differences across 184 countries leaves us with a limited set of variables (e.g. basic socioeconomic characteristics)" (1997, 20).

4. For a detailed analysis of the method used see Hofstede (1980). The data were collected from IBM employees covering seventy-two national subsidiaries, thirty-eight occupations, twenty languages, and at two points in time: 1968 and 1972. In total, there were more than 116,000 questionnaires with over 100 questions each.

5. The short-term vs. long-term orientation dimension was in fact discovered by Michael Bond, although it is presented with Hofstede's original four dimensions in many studies.

6. For a dimension-by-dimension description of the implications of Hofstede's national cultural constructs for interactive network diffusion see Maitland (1999).

7. For a more detailed discussion of culture and diffusion theory see Maitland (1999).

8. It is tempting to say "predictive elements" but Rogers contends that these categories do not provide predictive support, thus the term explanatory is used.

9. One reason for the lower than expected relationship may be the extent to which the computers are attached to networks. A measure of the existence of research networks was considered for the present study; however, the low number of observations for the proxy variable, government expenditures on R and D, excluded it from the analysis. For further information on use of the Internet, specifically electronic communication, in Switzerland, see the chapter in this volume by Lucienne Rey.

10. When computers are connected to a network (or the Internet) they are assigned an IP address. If a computer has a permanent connection it is given a static IP address. Computers that "dial in" for their Internet access are assigned a temporary or dynamic IP address. Computers with static IP addresses are referred to as "hosts." Hosts have been used as a measure of Internet diffusion by the U.S. government (see Anon. 1997) and the OECD (see OECD 1996; Paltridge 1996).

11. Although an addition of a host to the Internet does not reflect the use of the wide range of services the Internet offers, the study of adoption here is similar to diffusion studies performed for the telephone, which also is used in a variety of ways. Furthermore, as with other diffusion studies,

adoption says nothing about the extent of use. In this sense this study is similar to early diffusion studies of birth control, which were concerned with adoption rather than actual use.

12. For example, an examination of GDP and PC data from 1990 to 1994 shows the top four and six countries, respectively, are consistent throughout the period. Changes in ranks for other nations are typically limited to small increases or decreases, but are also difficult to determine due to missing data.

13. It must be noted, however, that direct comparison of the strength of these indicators is not possible from the analysis presented in Table 5 because each indicator is reported by a different subset of countries. Although the adjusted R^2 takes into account differences in sample sizes, it does not remedy the situation that the models are being tested on different groups of countries. This limitation is addressed in subsequent parts of the analysis.

14. A simple interpretation of the R^2 statistical measure is the amount of variance in the dependent variable explained by the independent variables. Here it can be described as the amount of variance in the START variable that can be explained by the variables Teledensity, Call Cost, and English Language Ability.

15. In a regression equation the standardized betas allow for the comparison of the predictive strength of the independent variables. The regression equation for model 1a is START = .279 NWSPAPR + −.344 INTCALL + .308 SCHLNROL.

16. $r_{START,GEMPWR.TELEDEN} = .177$ ($p < .1$) and $r_{START,TELEDEN.GEMPWR} = .5511$ ($p < .01$) and $r_{START,TELEDEN.GDP\_CAP} = .430$ ($p < .01$) and $r_{START,GEMPWR.GDP\_CAP} = .335$ ($p < .01$).

17. This method will inflate the adoption percentage, an input to the growth measure, for those countries with low teledensity rates. The result is reduced variance in the growth measure, however, the countries involved in this part of the analysis have relatively high teledensity rates so the impact should be minimal.

18. $r_{GROWTH,GEMPWR.PCPTHO95} = .237$ (n.s.) and $r_{GROWTH,PCPTHO95.GEMPWR} = .3746$ ($p < .1$).

19. To better understand this measure its bivariate correlation with the other centrality measures were examined. It was found to have an insignificant and low correlation with International Call Minutes, and moderate correlations with CENTRALITY and LINKS (r=.287, −.315, respectively). Its strongest correlations were with PCs (−.597), Teledensity (−.531), and Gender Empowerment (−.530), all significant.

References

1997. "The Framework for Global Electronic Commerce." Washington, DC. Clinton Administration.

Allen, D. 1988. "New Telecommunication Services: Network Externalities and Critical Mass." *Telecommunications Policy* 12(3): 257–71.

Blumenthal, M. S. 1997. "Unpredictable Certainty: The Internet and the Information Infrastructure." *Computer* 30(1): 50.

Dekimpe, M. G., P. M. Parker, and M. Savary. 1994. *Modeling Global Diffusion*. Fontainebleau, France: INSEAD.

Dekimpe, M. G., P. M. Parker, and M. Savary. 1997. *"Globalization": Modeling Technology Adoption Timing Across Countries*. Fontainbleu, France: INSEAD.

Eastlick, M. A. 1993. "Predictors of Videotex Adoption." *Journal of Direct Marketing* 7(3): 66–74.

Educational Testing Service. 1996. "TOEFL Test and Score Data Summary." Princeton, NJ: Educational Testing Service.

Gatignon, H., J. Eliashberg, and T. S. Robertson. 1989. "Modeling Multinational Diffusion Patterns: An Efficient Methodology." *Marketing Science* 8(3): 231–47.

Geertz, C. 1973. *The Interpretation of Cultures*. New York: Basic Books.

Hahne, B. 1997. *State of the Japanese Internet 1997: Such Distance Traveled, So Far to Go*. <http://ifrm.glocom.ac.jp/doc/hahne.html>

Herbig, P. A. 1994. *The Innovation Matrix: Culture and Structure Prerequisites to Innovation*. Westport, CT: Quorum Books.

Hofstede, G. 1980. *Culture's Consequences: International Differences in Work-Related Values*. London: Sage Publications.

———. 1997. *Cultures and Organizations: Software of the Mind*. New York: McGraw-Hill.

International Telecommunication Union (ITU). 1997. "World Telecommunication Indicators Database." Geneva: ITU. (see <http://www.itu.int/>)

Leiner, B. M., V. G. Cerf, D. D. Clark, R. E. Kahn, L. Kleinrock, D. C. Lynch, J. Postel, L. G. Roberts, and S. Wolff. 1998. *A Brief History of the Internet*. ISOC. (see <http://www.isoc.org/internet/history/brief.html>)

Lin, N., and G. Zaltman 1973. "Dimensions of Innovations." *Processes and Phenomena of Social Change*, ed. G. Zaltman, 93–116. New York: John Wiley and Sons.

Lotter, M. 1998. Personal Communication.

Maitland, C. 1999. "Global Diffusion of Interactive Networks: The Impact of Culture." *AI and Society* 13: 341–56.

Marriott, M. 1998. "The Web Reflects a Wider World." *New York Times*. 18 June.

Mody, B., C. Maitland, et al. 1997. "Jobs in Teleutopia: An International Comparison of Employment in Telecommunication Services." Annual meeting of The International Studies Association, June, Toronto, Canada.

OECD. 1996. "Information Infrastructure Convergence and Pricing: The Internet." Paris: OECD.

Paltridge, S. 1996. "How Competition Helps the Internet." *The OECD Observer* 201 (August/September): 25–27.

Parker, P. M. 1994. "Aggregate Diffusion Forecasting Models in Marketing: A Critical Review." *International Journal of Forecasting* 10: 353–80.

Persell, C. H. 1984. *Understanding Society: An Introduction to Sociology*. New York: Harper and Row.

Press, L. 1997. "Tracking the Global Diffusion of the Internet." *Association for Computing Machinery* 40(11): 11–17.

Rickard, J. 1995. "The Internet by the Numbers—9.1 Million Users Can't Be Wrong." *Boardwatch* 9(12). <http://www.boardwatch.com/mag/95/dec/bwm1.htm>.

Rogers, E. M. 1986. *Communication Technology*. New York: The Free Press.

———. 1995a. *Diffusion of Innovations*. New York: The Free Press.

———. 1995b. "Diffusion of Innovations: Modifications of a Model for Telecommunications." *Die Diffusion von Innovationen in der Telekommunikation*, eds. M.-W. Stoetzer and A. Mahler, 25–38. Berlin: Springer.

Rosman, A., and P. G. Rubel. 1995. *The Tapestry of Culture: An Introduction to Cultural Anthropology*. New York: McGraw-Hill.

Sun, S. L., and G. A. Barnett. 1994. "The International Telephone Network and Democratization." *Journal of the American Society for Information Science* 45(6): 411–21.

Tornatzky, L. G., and K. J. Klein. 1982. "Innovation Characteristics and Innovation Adoption Implementation: A Meta-Analysis of Findings." *IEEE Transactions on Engineering Management* 29(1): 28–44.

UNDP. 1996. *Human Development Report 1996*. New York: United Nations Development Programme (UNDP).

Valente, T. W. 1995. *Network Models of the Diffusion of Innovations*. Cresskill, NJ: Hampton Press.

World Bank. 1997. *World Bank Statistics*. Washington, DC. (see <http://www.worldbank.org/data/databytopic/>)

Appendix

Part of the Full Sample Correlation Matrix for Analysis with START

Table 1
Variables with Bivariate and Partial Correlations

Variables	Measures	Year	N	Expect *r* with START	*r* with START (1 tailed)	Partial *R* with START	*r* with GDP_CAP
Depend. Var.	START		213				
Economy							
Per Capita GDP	GDP_CAP	1994	153	+	.630**	CONTROL	1.00
Trade	TRADE	1994	147	+	−.100	−.184*	.067
Research Networks	EXPR_D	1992	19	+	.416*	.221	.412
	DEFENSE	1992	75	+	−.118	−.194	.051
Access to Mass Media	NWSPAPR	1992	154	+	.637**	.312**	.761**
Education	SCHLNROL	1990	110	+	.574**	.318**	.598**
	EDUCBUD	1990	75	+	.226*	.167	.155
Personal Computers	PCPTHO95	1995	87	+	.567**	.015	.891**
Infrastructure							
Teledensity	TELEDEN	1994	200	+	.593**	.099	.885**
Cost of Access	CLCOST94	1994	128	−	.017	−.282**	.353**
	LESDPC94	1994	99	+	.254**	−.675**	.847**
	COMPET	1997	84	+	.257**	.186	.182
Centrality	INTCALMN	1994	170	+	−.245**	−.457**	.167*
	INTCALL	1995	139	−	−.586**	−.436**	−.451**
	CENTRAL	1989	93	−	−.434**	−.226*	−.438**
	LINKS	1989	93	+	.467**	.197*	.536**
Peripherals	FAX1993	1993	105	+	.310**	−.842**	.914**
Culture							
	GEMPWR2	1996	92	+	.590**	.320**	.632**
	UAI	1980	50	−	−.176	−.009	−.270*
	TOEFL	1996	167	+	.530**	.442**	.327**

* Correlation is significant at the 0.05 level (2-tailed).
** Correlation is significant at the 0.01 level (2-tailed).

Reduced Sample Correlations for START

Table 2
Reduced Sample Corrrelation Matrix for START

		START	GDP_ CAP	NWS PAPR	SCHL NROL	INT CALMN	INT CALL	FAX 1993	TOEFL	TELE-DEN
START		1.000	.574	.565	.587	−.147	−.666	.471	.454	.667
	Sig. (2−tail)	.	.000	.000	.000	.286	.000	.000	.000	.000
GDP_CAP		.574	1.000	.704	.564	.400	−.608	.905	.438	.911
	Sig. (2−tail)	.000	.	.000	.000	.002	.000	.000	.001	.000
NWSPAPR		.565	.704	1.000	.597	.092	−.446	.729	.440	.762
	Sig. (2−tail)	.000	.000	.	.000	.503	.001	.000	.001	.000
SCHLNROL		.587	.564	.597	1.000	−.102	−.409	.497	.339	.720
	Sig. (2−tail)	.000	.000	.000	.	.458	.002	.000	.011	.000
INTCALMN		−.147	.400	.092	−.102	1.000	−.142	.325	−.086	.179
	Sig. (2−tail)	.286	.002	.503	.458	.	.301	.016	.531	.190
INTCALL		−.666	−.608	−.446	−.409	−.142	1.000	−.542	−.407	−.620
	Sig. (2−tail)	.000	.000	.001	.002	.301	.	.000	.002	.000
FAX1993		.471	.905	.729	.497	.325	−.542	1.000	.482	.852
	Sig. (2−tail)	.000	.000	.000	.000	.016	.000	.	.000	.000
TOEFL		.454	.438	.440	.339	−.086	−.407	.482	1.000	.449
	Sig. (2−tail)	.000	.001	.001	.011	.531	.002	.000	.	.001
TELEDEN		.667	.911	.762	.720	.179	−.620	.852	.449	1.000
	Sig. (2−tail)	.000	.000	.000	.000	.190	.000	.000	.001	.

a Listwise N=55

Table 3
Full Sample Correlation Matrix for GROWTH

		GROWTH	LESD PC94	EDUC BUD	TOEFL
GROWTH		1.000	.516	.554	.394
	Sig. (2-tail)	.	.010	.001	.007
	N	48	24	31	46
LESDPC94		.516	1.000	.624	.574
	Sig. (2-tail)	.010	.	.023	.003
	N	24	24	13	24
EDUCBUD		.554	.624	1.000	.396
	Sig. (2-tail)	.001	.023	.	.034.
	N	31	13	31	29
TOEFL		.394	.574	.396	1.000
	Sig. (2-tail)	.007	.003	.034	.
	N	46	24	29	46
PCPTHO95		.591	.850	.525	.316
	Sig. (2-tail)	.000	.000	.002	.036
	N	46	22	31	44
GEMPWR2		.569	.535	.650	.541
	Sig. (2-tail)	.000	.033	.000	.001
	N	35	16	25	33
UAI		−.556	−.653	−.411	−.238
	Sig. (2-tail)	.000	.004	.037	.162
	N	38	17	26	36
FAX1993		.418	.766	.392	.125
	Sig. (2-tail)	.005	.000	.035	.430
	N	44	21	29	42
GDP_CAP		.342	.842	.342	.258
	Sig. (2-tail)	.017	.000	.060	.084
	N	48	24	31	46
TELEDEN		.414	.685	.308	.256
	Sig. (2-tail)	.004	.000	.092	.089
	N	47	23	31	45
NWSPAPR		.434	.443	.188	.108.
	Sig. (2-tail)	.003	.039	.320	.497
	N	44	22	30	42

* Correlation is significant at the 0.05 level (2-tailed).

** Correlation is significant at the 0.01 level (2-tailed).

(cont.)

Table 3 *(cont.)*
Full Sample Correlation Matrix for GROWTH

PCPT HO95	GEMP WR2	UAI	FAX 1993	GDP_ CAP	TELE- DEN	NWS- PAPR
.591	.569	−.556	.418	.342	.414	.434
.000	.000	.000	.005	.017	.004	.003
46	35	38	44	48	47	44
.850	.535	−.653	.766	.842	.685	.443
.000	.033	.004	.000	.000	.000	.039
22	16	17	21	24	23	22
.525	.650	−.411	.392	.342	.308	.188
002	.000	.037	.035	.060	.092	.320
31	25	26	29	31	31	30
.316	.541	−.238	.125	.258	.256	.108
.036	.001	.162	.430	.084	.089	.497
44	33	36	42	46	45	42
1.000	.685	−.394	.813	.853	.813	.544
.	.000	.016	.000	.000	.000	.000
46	34	37	43	46	45	42
.685	1.000	−.282	.604	.611	.649	.483
.000	.	.111	.000	.000	.000	.004
34	35	33	35	35	35	33
−.394	−.282	1.000	−.411	−.220	−.207	−.296
.016	.111	.	.011	.184	.218	.084
37	33	38	37	38	37	35
.813	.604	−.411	1.000	.889	.808	.722
.000	.000	.011	.	.000	.000	.000
43	35	37	44	44	44	41
.853	.611	−.220	.889	1.000	.854	.644
.000	.000	.184	.000	.	.000	.000
46	35	38	44	48	47	44
.813	.649	−.207	.808	.854	1.000	.615
.000	.000	.218	.000	.000	.	.000
45	35	37	44	47	47	44
544	.483	−.296	.722	.644	.615	1.000
.000	.004	.084	.000	.000	.000	.
42	33	35	41	44	44	44

Table 4
Reduced Sample Correlations

		GROWTH	START	TOEFL	PCPTH O95	GEMP WR2
GROWTH		1.000	.132	.388	.614	.562
	Sig. (2-tail)	.	.504	.041	.001	.002
START		.132	1.000	.342	.549	.452
	Sig. (2-tail)	.504	.	.075	.003	.016
TOEFL		.388	.342	1.000	.341	.526
	Sig. (2-tail)	.041	.075	.	.075	.004
PCPTHO95		.614	.549	.341	1.000	.697
	Sig. (2-tail)	.001	.003	.075	.	.000
GEMPWR2		.562	.452	.526	.697	1.000
	Sig. (2-tail)	.002	.016	.004	.000	.
UAI		−.561	.131	−.361	−.385	−.271
	Sig. (2-tail)	.002	.508	.059	.043	.163
FAX1993		.508	.521	.270	.843	.663
	Sig. (2-tail)	.006	.004	.165	.000	.000
GDP_CAP		.449	.579	.328	.864	.663
	Sig. (2-tail)	.017	.001	.088	.000	.000
TELEDEN		.498	.648	.367	.861	.694
	Sig. (2-tail)	.007	.000	.055	.000	.000
NWSPAPR		.479	.518	.200	.685	.488
	Sig. (2-tail)	.010	.005	.309	.000	.008

Listwise N=28

Table 4 *(cont.)*
Reduced Sample Correlations

UAI	FAX 1993	GDP_ CAP	TELE-DEN	NWS-PAPR
−.561	.508	.449	.498	.479
.002	.006	.017	.007	.010
.131	.521	.579	.648	.518
.508	.004	.001	.000	.005
−.361	.270	.328	.367	.200
.059	.165	.088	.055	.309
−.385	.843	.864	.861	.685
.043	.000	.000	.000	.000
−.271	.663	.663	.694	.488
.163	.000	.000	.000	.008
1.000	−.339	−.199	−.174	−.180
.	.078	.310	.376	.359
−.339	1.000	.914	.824	.797
.078	.	.000	.000	.000
−.199	.914	1.000	.867	.807
.310	.000	.	.000	.000
−.174	.824	.867	1.000	.753
.376	.000	.000	.	.000
−.180	.797	.807	.753	1.000
.359	.000	.000	.000	.

Countries in the 'START' Sub-Sample Analysis

Australia	Finland	Mauritius	Sierra Leone
Austria	Ghana	Mexico	South Africa
Bangladesh	Greece	Mongolia	Spain
Belgium	Hungary	Morocco	Sri Lanka
Benin	India	Nepal	Sweden
Brazil	Iran	Netherlands	Switzerland
Burundi	Israel	Norway	Tunisia
Canada	Italy	Pakistan	Turkey
Chile	Kenya	Peru	United Kingdom
Costa Rica	Korea	Philippines	United States
Czech Republic	Latvia	Poland	Uruguay
Denmark	Lithuania	Portugal	Venezuela
Ecuador	Malawi	Russia	Zambia
Egypt	Malaysia	Saudi Arabia	

Countries in the 'GROWTH' Sub-Sample Analysis

Australia	Ecuador	Mexico	Spain
Austria	Finland	Netherlands	Sweden
Belgium	Greece	Norway	Switzerland
Brazil	India	Philippines	Thailand
Canada	Japan	Portugal	United Kingdom
Chile	Korea	Singapore	United States
Denmark	Malaysia	South Africa	Venezuela

Countries in the 'GROWTH' Full Sample Analysis

Argentina	Ecuador	Italy	Portugal
Australia	Estonia	Japan	Singapore
Austria	Finland	Korea	Slovakia
Belgium	France	Latvia	South Africa
Brazil	Germany	Luxembourg	Spain
Bulgaria	Greece	Malaysia	Sweden
Canada	Hong Kong	Mexico	Switzerland
Chile	Hungary	Netherlands	Taiwan
Costa Rica	Iceland	New Zealand	Thailand
Cyprus	India	Norway	United Kingdom
Czech Republic	Ireland	Philippines	United States
Denmark	Israel	Poland	Venezuela

II. Theory/*Praxis*

New Kids on the Net: Deutschsprachige Philosophie elektronisch

Herbert Hrachovec

> The old, albeit hackneyed, computer expression "GIGO"—Garbage In, Garbage Out—has been removed from vocabulary and rhetoric at a time when it seems most needed. The hype about the Internet has in fact created a new enchantment in Western societies. Dealing with the realities of "virtual reality," however, will be a process of progressive disenchantment wherein the limits of communication and information as the essence of emancipation become clear. The Net, then, has attained a status much like God . . . before rationalisation.
>
> —*Interrogate the Internet*

The Internet protocols offer several modes of global, digital data transfer by procedures like telnet, ftp (File Transfer Protocol) or SMTP (Simple Mail Transfer Protocol).[1] Some modes are designed to enable exchange of information between single users or to allow access to remote operating systems. There are, on the other hand, a number of techniques specifically developed to support social interaction: "Chats" (Internet Relay Channels) or "MUDs" (Multi-User Dimensions). Mailing lists fall somewhere in between those two categories, basically building on the person-to-person SMTP, but enhancing it (often by extensive use of mail aliases) to establish electronic discussion groups. Discourse on such lists is generally more civil and substantive than on Usenet, but still considerably more chaotic than any traditional form of written public exchange. While chatters may open or close new "channels" at will and participants in Usenet's alt-hierarchy indulge in their freedom to create and discard any number of quixotic newsgroups, list-owners need some administrative support to install and configure the necessary software which makes for a comparatively stable, restrained communicative environment.

My topic will be quite specific, namely an overview of German-language mailing lists in philosophy. The purpose of the discussion is, however, a more general one: to explore the tension inherent in implementing a tool for global communication in a very particular geographical and professional context. As a preliminary, let me briefly name the lists respectively and add some general remarks on e-mail in a global context.

Give-l, which ran from December 1994 to September 1996[2], was the first attempt to establish an electronic discussion forum for German-speaking philosophers on the Internet and it exhibited a slightly half-baked enthusiasm I will comment upon shortly. Eventually *give-l* could not contain the contradictions between its naive universalism and its de facto clientele. A more discriminating approach seemed to be called for. My second focus will be the story of *real*, an e-mail forum intended to support lecture courses I gave at the Department of Philosophy at Vienna University starting in fall 1996.[3]

Methodological reflection had by this time set in and I shall report the consequences of a more sober approach to the technological challenge. *Give-l* was a success while it lasted, *real* was sometimes lively, but very often sluggish and in constant need of prompting. These difficulties encountered with *real* will lead to a discussion of the inevitable disenchantment with de-contextualized, but necessarily local implementations of global communication software. A more pragmatic approach suggests itself. My third example will be *philweb*, a Hamburg-based list that has been very active recently.[4] The vast majority of its members are students of philosophy at various German universities. *Philweb* is a second-generation mailing list, sometimes containing echos of foundational moments, but more often busily exploring the newly discovered opportunities.

This talk will be a small-scale *Bildungsroman* starting with the blissful coincidence of the general and the particular and eventually leading to a more detached assessment of the prospects of an initial synthesis of technology and culture. But before I begin to relate my story, some reminders concerning the overall framework of Internet communication might be helpful.

Mailing lists tend to be shaped by core groups of dedicated participants, developing their interests and opinions in front of a predominantly receptive audience of subscribers. A new kind of communicative *praxis* is established on top of some guidelines on how computers should exchange data: participation in quasi-instantaneous, globally distributed, non-hierarchical discursive interchange. Computer networks, as is well known, are not confined by any historical or geographical borders. As a consequence, the cultural impact of the technical devices seems to affect arbitrary collections of users who avail themselves of the necessary equipment and know-how. One of the most dazzling experiences of communication on the Net, it has correctly been pointed out, is its global egalitarianism. While it is true that large parts of the planet are still excluded and the predominance of the English language imposes important constraints on the participants, it is difficult to avoid an initial euphoria, a cosmopolitan state of mind, as one becomes familiar with a machinery that can support spatially unlimited cooperation between equals with a minimum of administrative overhead.

The rules of TCP/IP have been laid down in one country, at a particular time, under particular circumstances, but the scope of their application is universal. Their inherent capacity to transform information exchange all over the world seems much more powerful than any special pleading in favour of local sensitivities. This way of looking at the Internet is, obviously, reminiscent of well-known philosophical debates centering on the universality of Eurocentric Reason. There is a tension, if not a paradox, in one country determining the address space for all of the world. Hegemonical attitudes are very much in evidence as the participants—government, big business and transnational agencies—struggle for authority and their share of bandwidth. Appeals to "international standards" are often quite partial. But it is equally important to realize that nobody forced the Internet on the non-US part of the globe. The universalised rules of TCP/IP are acknowledged and, indeed, put to use, by numerous local communities drawing profit from international standards they have not, admittedly, been asked about. So, here is an account of how one such activity developed.

give-l

The designation "*give-l*" and the original purpose of the list are in themselves indications of the tension I have indicated. The acronym was supposed to stand for "Globally Integrated Village Environment," referring to a local Viennese research project trying to put Marshall McLuhan's ideas to the test. The list was established to support the activities of the research team and I spent some of my seminars discussing their agenda. The result was a strange mix between universal reach and local circumstances.[5] Several scholars, searching the net for keywords like "global" and "village" were in due course directed to *give-l*—only to be disappointed when they discovered that German was the dominant language on the list. English was also acceptable and was indeed used by some participants feeling more comfortable in their native language. Reading German was, however, a prerequisite of actively participating, a fact that had simply been overlooked when the acronym was chosen to attract an international audience.

It took list members several month to become aware of this dilemma and some more time until a new reading of give was proposed: "Gehirne in vollem Einsatz" (roughly "Brains giving their best"). This playful echo of the original meaning of the list's name did not, however, remove a more fundamental ambivalence acutely felt at the time. Viennese students were suddenly exchanging their opinions and pursuing their academic curriculum in front of a worldwide audience. Describing the situation in these terms might sound unduly pathetic. Still, I want to argue that the description is—up to a certain degree—legitimate. Compare the thrill of suddenly talking to ten thousand people over a microphone. An individual voice is suddenly broadcast by an enormously powerful medium. To disregard the phantasies such scenarios evoke makes for a severely restricted philosophy.

Foundational experiences are not for keeping, but neither are they just discardable by-products as history unfolds. Starting January 1995 a lot of traffic on *give-l* was concerned with administrative troubles as well as with several papers written on the occasion of a symposium sponsored by the City of Vienna. But there was a less pragmatic undercurrent: no one had done this kind of thing before.[6] Some (largely implicit) account of what the activity amounted to was presupposed in our practice. In the background of computer-mediated transactions a proto-theory of mailing lists was taking shape.

I was, as it happened, at that time commuting between Essen, Germany and Vienna, using the list for some teleteaching. The list itself eventually included about one hundred fifty persons of which approximately fifty were based in Vienna, often knowing one another personally, e.g., from taking part in my seminars. Under these circumstances a certain technologically induced euphoria took hold of several contributors. It has often been remarked that e-mail combines features of writing and conversation, producing "texts" that carry some of the immediacy of face-to-face encounters. This feature was certainly appreciated, but another, more conceptual peculiarity of e-mail discourse impressed itself even more deeply on the group. Texts (or tele-events), when broadcasted all over the world, often produce an inherently passive audience that has no choice but to accept whatever the distributors make available. Local meetings, seminars for example, provide opportunities to shape events in person. Technically speaking, mailing lists are trivial extensions of SMTP, but they offer entirely new social dynamics.

The notion of a "global audience" has in the past, somewhat metaphorically, been applied to people reading their daily paper or sitting in front of television sets. With the invention of mailing lists the term can be given a much more literal meaning. Real-life audiences are distinguished from "audiences" in a derived sense by their member's actual awareness of each other. Public events in their most basic form demand bodily presence and enable people to react to each other's interventions spontaneously, whereas a media event synthesises numerous single addressees into a more abstract social gathering. The mechanism of mailing lists, as it turns out, goes a long way to combine the requirements of global reach and local awareness. One might be able to watch one's neighbours watching TV, or notice the book one's friend just bought, but there is no way to know in general who at a given moment is watching a particular program or what persons are reading one's favourite book.

In contrast to this, every mailing list has a simple "review" command, enabling each member to automatically retrieve the names of all fellow-participants. This is, admittedly, not the bodily co-presence characteristic of on-location meetings, but it is one of its closest approximations yet by means of media technology. Participants in mailing lists de facto know precisely whom they are addressing themselves to and they know that those addressed know that they are noticed in this way. Furthermore, if the system works, electronic mail is practically simultaneous on a global scale, so that responses to a message can in principle be given in real time. A group of people might be dispersed all over the planet and still each of its members can know of each other, address the group at any time and receive instant feedback, which is itself subject to quasi-immediate comment. As these possibilities dawned on some of the members of *give-l*, exchanges on the list acquired an importance far exceeding the issues at hand.

For a time it seemed that one could have the best of two worlds: instantaneous social interaction without bodily presence.[7] Key members knew each other and physically met: still they were thrilled by the opportunity to communicate via e-mail messages, sometimes sitting next to each other in the computer lab. Their real-life existence had somehow acquired an electronic supplement as their identity as participants on *give-l* exerted increasing influence on their actual life. I had loosely associated *give-l* with a seminar I held at the Department of Philosophy expecting it to enhance traditional forms of learning/teaching. But the list quickly developed into a melange of discussions only temporarily focused on single topics. High-quality contributions were running side by side with beginner's questions and silly comments, mirroring a student's checkered experience at an academic institution in a way conventional media are unable to match.

Inevitably, as a group identity was forged, a social hierarchy imposed itself on the participants.[8] This lead to predictable tensions on-line and in real life. One list member, to mention the most controversial case, intermittently attacked his fellows quite rudely, even though he could be seen as a reasonably well-mannered, if idiosyncratic, student in the context of the seminar meetings. Knowing this person's peculiarities, a majority was prepared to tolerate his transgressions on the list. But when newcomers from outside the local circle were also fiercely attacked the affair threatened to get out of hand and, after several warnings, I removed the offender from the list.

The consequences of this removal were dramatic and served as a first reminder of the more problematic aspects of on-line meetings. Two weeks after the event a student, resenting my decision, asked "whether all *give-l* members are fascists?" This provocative question shattered the—until now, largely innocent—preconception of a more productive, civil life in cyberspace, leading to a bitter flame war among several proponents. On reflection the reasons for this nasty confrontation turn out to be closely connected to the possibilities praised in my previous remarks. The questioner, actually a rather withdrawn, courteous person, was simply unaware of the impact a single word could have in an environment that carries no collateral information on the personal bearing and attitude of the speaker/writer. This sort of disembodiment is quite possibly a remedy against stifling prejudice, but it can also severely disturb social interaction.[9]

One ambivalent phrase, not embedded within the usual context of situated know-how, dropped into a digitally enhanced community, can trigger a completely unforseen chain of reactions, possibly leading to the self-destruction of the group. Electronic communities are (somewhat miraculously) built upon transmission techniques and words alone—and can just as easily be destroyed by hardware failure or a single inappropriate utterance. Luckily, *give-l* survived this crisis and continued to provide a learning environment for many of its participants. When, for example, teachers and students at the University of Vienna went on strike against severe budget cuts proposed by the Austrian government in spring 1997, *give-l* featured some excellent conceptual and economic background information as well as extensive discussion of the options facing the academic community.[10] Yet, after having run for over three semesters, the list showed distinct signs of wear.

real

At the establishment of *give-l*, all its members had shared a certain amount of curiosity and a fair measure of ignorance regarding the whole enterprise. As the list developed, this background obviously changed. At the beginning the very fact of "being connected" was felt to be of overwhelming importance and mutual encouragement was as welcome as carefully prepared arguments. But the pursuit of academic learning and indulgence in the unconstrained voicing of opinions do not easily fit together. There were some attempts to impose a more conventional structure upon the discussion, all of which failed. Mailing lists—rather like lively meetings of friends—do not easily allow for this kind of administrative regulation. As a consequence, contributors who had spent considerable energy in setting up a philosophical discourse gradually grew disenchanted, unwilling to deal with the concurrent "gossip" on a daily basis. With the original excitement subsiding, a different arrangement was decided between the Viennese proponents of *give-l*.

The list was to be split in two, one part retaining the "brand name" and offering a club-like atmosphere for students at the department, whereas the other part was meant to supplement my Viennese teaching, carrying theoretical discussions exclusively. The new list *give*, I am sorry to report, proved an instant failure. The special mix of personalities and mechanical gadgets that had produced and supported *give-l* could not be duplicated in the quickly changing area of digital technologies. The second list, *real*, proved more enduring. It took its name from the lecture course it was to support—"Wirklich, möglich, virtuell" ["Really, possibly, virtually"]—but there was also a hint at the list being more realistic regarding the possible functions of electronic discourse. Still, with a lot of interest in teleteaching and experimental use of the new media, expectations were high.

"Virtuality" is an intriguing concept and *real* started with a prolonged discussion of how digitalised representation should be distinguished from "reality" and "possibility." The spectrum of contributions was fairly broad, ranging from physics to postmodern theory and self-referential comments on the "virtual" nature of the list itself. Cooperative philosophical explorations seemed to be possible within this framework. But when the topic of "virtuality," after two months' time, had lost its attraction, the list could not maintain its initial momentum. It did never, in particular, produce the kind of group-consciousness that had been a hallmark of *give-l*.

The highlights of *real* occurred when—for some generally unpredictable reason—an issue or an event caught the imagination of several participants, leading to a short, intensive exchange which usually broke off as abruptly as it had begun. And when I tried to repeat my attempts at teleteaching, arranging for two groups of students from Vienna and Weimar to share the list for mutual comments on lectures I had given in both cities, the proposal did not meet with any significant interest. Mailing lists are, according to this experience, of only limited use in supporting comparatively high-focused academic cooperation. This seems to be the opposite side of their very informality. It is precisely because they enable people to react to other people's interventions quickly and spontaneously that they do not easily provide an environment conducive to doing "serious" philosophy.

My notions of seriousness can, of course, be challenged at this point. A certain species of "media philosophy" is intent on explicitly rejecting the traditional professional standards that I am implicitly invoking here.[11] According to their pronouncements, future philosophical efforts should make the most of multimedia, hypertextualized technology, breaking free from the confines of one-step-after-the-other linear argument. I do not deny the attractions of those manifestos and tend to follow their advice—once in a while. But I am not prepared to overlook the severe limitations imposed on academic endeavours by technologically-mediated, unconstrained exchange of opinions.

Mailing lists are a valuable tool as long as having an equal voice and communicating with a minimum of administrative hassle are the most important requirements. It is not impossible to employ them for bona fide educational purposes like tutorial guidance or careful slow readings of classical texts. Yet, the inherent egalitarianism of the procedural substratum of mail aliasing seems to be somewhat at cross-purposes with attempts to build the stable, mildly hierarchical structures known from ordinary teaching. Precisely because the usual framework of time and space is drastically altered and physical presence replaced by written communication, the metaphor of an "electronic classroom" is of limited use. The hesitant conclusion from running *real* is, therefore, that it is probably a mistake to expect much philosophical content even from special-purpose mailing lists. Since this is a somewhat negative result the question of its relevance to the vision of a global, unrestricted, well-informed exchange of ideas naturally arises.

Questioning students about their reluctance to involve themselves with *real* produced some straightforward, pragmatic reasons for the partially disappointing developments. In 1994/95, the World Wide Web had not yet achieved the overwhelming importance it was to reach by the second part of 1996 when *real* was started. To students fascinated by links, graphics and animation, simple e-mail seemed somewhat austere and could not capture the imagination to the extent necessary to engage in prolonged philosophical dialogue. Confronted with a seemingly unbounded supply of intellectual freeware most users found it increasingly difficult to concentrate on complicated issues when on-line. The omnipresence of web-browsers, most of them including e-mail functionality, overshadowed the notion of a mailing list which does not, after all, offer anonymous surfing to the general public. Putting *real* on the Web did not, incidentally, help. Hyper-mail is helpful in making technical support accessible or in simply sharing some information with a broad audience. It is not, for this very reason, well-suited to the purposes I tried to put it to.[12] Such are the risks one has to reckon with when entering unexplored territory. But there is a more substantial philosophical lesson to be drawn from reflecting on the development of *give-l* and *real*.

In comparing the two lists, some of the enthusiasm surrounding *give-l* can be seen from a different perspective. I have hinted at the ambivalent nature of exempting the body from what is otherwise a characteristically communicative setting. This holds for mailing lists (or chats and MUDs) in general. There is, however, an additional aspect unique to foundational moments in global electronic communication. When first confronted with a technical tool like the Listserv software an almost automatic reaction is to run together two different projections, namely the procedural advantages of the technology and its perceived usefulness to the particular situation one finds oneself in. Such technologies—at a first encounter—present themselves as a hybrid between context-independent promises and very specific expectations. Typing at her keyboard, a person can reach a global audience. I am not denigrating this hybrid form. It seems to me that, on the contrary, its power has to be acknowledged and its presuppositions have to be scrutinised.

One might, tentatively, say that an imaginatory cross-fertilisation is at issue here. The rules of SMTP contain nothing to inspire widespread phantasies, whereas the phantasy of all the inhabitants of the planet communicating unrestrictedly has probably been around for as long as humanity itself. Inconspicuous moments like making an appointment at the computer lab, determining the parameters of a mailing list's configuration files, etc., can, surprisingly, acquire pivotal importance by short-circuiting technological capacity and an external content that is imaginatively superimposed upon the working of the machinery. This is not, to repeat my point, meant to be a deconstruction of such incidents. Rather, examining their inherent structure we learn about the force and the limits of attempts to install a computer-mediated space of Reason.

It is tempting to put the point in Hegelian terms: mailing lists exhibit the principle of widely-distributed, democratic, simultaneous discourse *an sich*, i.e., formally, by virtue of their technical definition. The corresponding philosophical notions remain, on the other hand, *für sich*, confined within the realm of theoretical design. In order for the promise to work itself out, both sides would have to be mediated, exploring the power of operational, but abstract procedures to shape and transform imagination via actual discourse. This, of course, is where the hybrid construction is put to a test it cannot possibly pass. Philosophical talk of rationality, generality and social symmetry is not meant to be taken in the literal sense a mailing list exemplifies. Some enthusiasts, it is true, start off with a simplistic understanding of terms like "universality" and "immaterial"; their punishment consists in having to deliver papers tracing their disenchantment. Yet, as Wolf Biermann, a German songwriter, put it in a different context: "*Wer sich nicht in Gefahr begibt, der kommt drin um.*" Not taking risks is living dangerously.

To mention a similar dilemma, it is, at a first glance, a very plausible proposition that Roland Barthes and Jacques Derrida (among others) are prophets of digitalised hypertext which neatly materialises their conceptual design (Landow 1994). But, taking a closer look, it becomes obvious that the architecture of a book like Roland Barthes' S/Z is completely foreign to the current realities of hypertext. Writing about "nodes" and "networks" in a traditional context is importantly different from designing HTML pages and similarities between these two activities are extremely superficial. The meanings of the term "global" in the parlance of media theorists and philosophers are, likewise, related by family resemblance—at best.

The general topic of this conference is the impact of globally distributed technologies on local communities shaped by history and custom. Some suggestions emerge from the preceding discussion.

With the benefit of hindsight it is comparatively easy to find a familiar pattern in my account of *give-l* and *real*. Life is not more enlightened since electricity is generally available and foreign countries are not necessarily better known to us since we can get there by plane. Continental philosophers have warned us all along against being fooled by formalisms devoid of content and even software designers are beginning to inquire after the needs of particular users before implementing their programs (Winograd, 1996). It seems to follow that the entire procedure—establish a mailing list, ask questions later—was misguided, a typical example of falling prey to mere appearances. I do not want to dismiss the charge out of hand and I certainly concede that I'd do things differently the second time. Yet, such more cautious approaches are themselves built on presuppositions that are at least as dubious as the myth of empowerment by mere technology.

Conventional wisdom has it that there is a realm of science and technology which holds great promise for mankind, even though it is simultaneously perceived to be a dangerous force, quite likely to trigger enormous devastation. In order to check the techno-experts we need prudence, the power of good judgement, the humanities. This is because history and the social sciences teach us about the constraints every society and every cultural environment imposes on the machinery it needs for its survival. But notice the dualism deeply entrenched in this point of view.

The strategic recourse to the powers of the mind is, it seems to me, just as problematic as unguarded technophilia. In preserving a domain of detached reflexion it simultaneously renders technology immune against any direct intervention. "Humanists" are not supposed to meddle with the formalism, their area of competence being the scholarly assessment of its possible consequences. This attitude, I suggest, does not do justice to the way technological achievements capture our imagination and tempt us to explore their potential. Running a mailing list in the early days of the Internet is a permanent transgression, challenging many established rules of behaviour and provoking questions that have never been asked before. But this is the subject matter of a different talk, so I will conclude this section with a one-sentence indication of my personal position: philosophy disposes of an enormous amount of knowledge, some of which can well be put to unauthorised use by newcomers and even dilettantes as they take up a challenge previously unknown.

philweb

What I've been saying amounts to an extended answer to the following question that was put on several mailing lists dealing with philosophical topics on February 14, 1998:

> I wonder what are the main email lists for philosophical discussions. I am not looking for a specific topic, but philosophy in general. By main lists I mean lists where the discussion includes all kind of philosophies, as well as reference to what is going on today in the area.

As Jim Morrison was singing in the late sixties: "We want the world and we want it now." This is not going to work, but it is not completely crazy either. I was surprised at the courtesy with which this inquiry was met, the sender simply being referred to some of the well-known listings of philosophical resources. On closer inspection, though, simple-minded interventions like the question quoted above raise more interesting issues. What are we to expect from the ubiquity of such naïve enquiries? Can mailing lists overcome the constant danger of being deflated? Can philosophical activity be adjusted to profit from the potential of permanent ad hoc disturbance?

One possible reaction is to settle for administrative information. Philos-L offers professional services to English-language philosophers and I have established a similar list (*register*) to serve the academic community in German-speaking countries.[13] But such undertakings, while clearly being useful, provide a very limited answer to the general worry. Electronically addressing the members of the profession is highly convenient and will undoubtedly become even more widespread in the future—but what about content? Will it be affected by its means of proliferation? It should by now be obvious that putting the issue in such general terms will only provide utopian (or dystopian) guesswork. The question's scope has to be restricted and I will base a tentative answer on my familiarity with the current employment of the Internet for philosophical purposes in Austria, Germany and Switzerland.

Increasing numbers of German-language universities are present on the Web, offering the usual set of information, including brief overviews of their departments of philosophy. There are approximately seventy home pages of philosophy professors, most of them embedded within the general presentation of their institution. Less than twenty of those home pages contain more than a CV, a list of publications and a description of past and current interests. Some philosophical associations like the "Ludwig Wittgenstein Gesellschaft" or the "Austrian Society for Philosophy" are on-line and a number of publishing houses as well as academic journals supply electronic catalogues and indices. All of this pretty much mirrors the US-American situation, albeit on a smaller scale. But, turning the attention to cooperative projects, there are interesting differences.

With the exception of Vienna University, up to now there have been next to no attempts to take up the challenge of computer-mediated philosophy in an institutionalised, academic context. German philosophy departments tend to be quite hierarchically organised, unwilling and unable to quickly adapt to outside pressures and public expectation. On a more conceptual level, most of the established theoretical frameworks profess a distance towards mass media and the marketplace of ideas. Experimental electronic philosophy is, consequently, done by a small group of graduate students and people on the fringe of the educational system. The authoritative collection of digital resources in German-language philosophy is maintained by Dieter Köhler, a graduate student from Heidelberg, in his spare time[14] and one of the most charming sites, "Annette's Philosophenstübchen" is an open attempt to challenge the kind of philosophy usually done in academia.[15] Probably Germany's most noteworthy contributions to on-line life in philosophy have been provided by *PhilNet*, a small group of students very loosely affiliated with Hamburg University.

I'll restrict myself to the mailing list initiated by the Hamburg group in May 1996, incidently on the very same day that I launched *register*. After some initial confusions the list-owners reached an agreement concerning the respective profiles of their lists. *Philweb* was to cater for net-users and web-designers interested in applying new information technology to the field of philosophy. These aims were in line with several other *PhilNet* activities, such as building a philosophical search engine and a text repository. The project had difficulties in developing: there were few responses and traffic on *philweb* had virtually stopped when (in September 1997) the list suddenly exploded.

Two or three philosophy professors, several (graduate) students and some extra-academic participants had locked into intensive discussions and were producing considerable output on issues as diverse as "Realism and Anti-realism," "Consciousness," "Colours and Sounds," "Goethe," and "Bombing Iraq." This was not, I hasten to add, Usenet material, but more often than not carefully developed arguments taking note of other people's view, civil and enterprising at the same time. The spirit of the list can probably be best compared to that of "Philosophy and Literature," a list run at the University of Texas. But *philweb* had negligible institutional support and no pre-set agenda to begin with.

There is a certain irony in the fact that Georg Sommer, the spokesman of *philweb*, had not envisaged this type of philosophical discussion and had, in fact, withdrawn from the list at the time it was more or less reinvented in a new format. It took some administrative lacunae for the participants to realize that the list's owner was not even a member of the list any more. He had to be re-invited to give his opinion on recent developments. An understanding was quickly reached: list ownership passed to two of the participants and it was generally agreed to continue the list as a forum of prolonged philosophical brainstorming.

Free electronic discourse follows its own somewhat impredictable laws and my guess is that *philweb* will not be able to maintain the impressive quality it had reached at the beginning of 1998. In this instance, as in the case of *give-l*, a surprising amount of cognitive energy was in evidence, strangely fused with excitement concerning technologies conveniently supplied by a computer lab. For an initial stretch of time philosophical activity, generously shared among the group, is oblivious to doctrines, curricula and grades. *Philweb*'s success will quite possibly be short-lived—but what kind of attitude is at work in such predictions? Mailing lists are, after all, neither hardcover publications nor traditional social structures. The new kids articulating themselves on *philweb* should not be submitted to a set of criteria taken from quite different institutionalised settings. They will probably fail to get credits for their efforts, but their experiments in establishing a transitory, digitally distributed verbal agora cannot fail to affect the future of philosophical scholarship.

The feasibility of quasi-instantaneous, two-way global data transfer in a public medium evokes, as all of you know, hopes of increasing democratic participation among citizens and within various organisations.[16] As this miniature *Bildungsroman* draws to a close, one of its lessons is that, unfortunately, at this level of generality the desirable effects of each participant having an equal voice and basically similar chances to contribute to a common goal can not be separated from the nightmare of computer-mediated witch hunts. Involvement in mailing lists similarly suggests that their procedural advantages, compared to traditional communication, can be a dubious blessing, provoking exalted expectations and impeding a sober analysis of how the new media might affect the Humanities. I have specified a more restricted terrain to begin to answer the question of the Internet's implications for philosophy. Scholarly work is, on the one hand, fairly rigidly determined by professional standards while, on the other hand, often characterised by a spirit of tolerance and mutual respect. Even though both *give-l* and *philweb* shared some of these qualities they were not their most important contributions to the issue at hand.

By shifting the ground from the classical manipulation of texts towards instantaneous textual publicity, people writing on these lists changed some basic rules of literacy. Rather than being presented in curricular modules, philosophy could be seen as a continuous group-activity, permeating the week in between classes, blending local settings and external interventions. Rather than following given institutional patterns such activities could arise (and disappear) spontaneously, uncoerced by efficiency testing and financial constraints. Such lists, to summarise, produce a new genre: semi-scholarly on-the-spot writing, transmissible across the planet. I did not, in this talk, present examples of how serious (or how annoying) electronic philosophical discussion can get at close view. Suffice it to say that the list's archives have been indexed by the big search engines and that the log files show considerable interest in many of the issues discussed over the years. This is another prospect of things to come: continuous digital availability of day-to-day discourse. (I'll not pass judgement on whether this is a good thing or a nuisance.)

None of this will change the merits of a single philosophical argument, but it might well contribute to shifting the ground on which traditional philosophy itself rests. General principles and universal rules have always been prominent concerns for philosophers, even while their means of communication were quite specific: books, papers, lectures. This discursive frame has not been seriously challenged by the advent of mass media and one-way broadcasting. Neither the telephone nor TV has had any tangible impact on the way philosophy is done. There is a chance that the constitution of a permanent, communicative, electronic space and the development of virtual philosophical communities within this space will be of greater importance. Exchanging texts and arguments on an equal footing is, after all, an elementary philosophical gesture which will be heavily affected by the possibilities opened up by the Internet.

I have not hidden my ambivalence concerning the promises of a digital wonderland. Reviewing the dynamics of three mailing lists allows the reasons for a skeptical attitude to emerge more clearly. Some features of the new discursive forms are incompatible with the current educational system. Expecting strictly focused discussion within a twenty-four hour show is bound to prove disappointing. There is, on the other hand, no way to beat mailing lists when it comes to addressing a world-wide audience and (albeit in a rather specific sense) implementing the principles of universality often discussed in philosophical treatises. Theoretical activities have suddenly become available within the framework of a mass medium, and it is far from clear how this encounter is going to work out. The Net is not the most natural habitat for German-language philosophers. It is, in fact, yet undecided who its typical inhabitants will turn out to be. In the meantime, most are new kids, sporadically at unease and frequently sounding strange.

Notes

This manuscript appeared originally in the *Electronic Journal of Communication / La revue electronique de communication*, 8 (3 & 4), 1998 (see <http://www.cios.org/www/ejcrec2.htm>) and is reprinted by kind permission of the editors.

1. For technical information see Tanenbaum (1996). The motto is taken from Shields (1996, 131).

2. The list is archived at <http://hhobel.phl.univie.ac.at/gl>. Andreas Krier, Oliver Marchart, Gabriele Resl, Horst Tellioglu and Monika Wunderer have been most helpful in making *give-l* an exciting place. Thanks to all of them.

3. Cf. <http://hhobel.phl.univie.ac.at/real/realarch>.

4. For information see <http://www.sozialwiss.uni-hamburg.de/phil/ag/philweb.html>.

5. Mitchell (1995, 6–24) includes a fine phenomenological description of this feature of electronic agoras.

6. For an overview of the general principles of digital socialisation see Baym (1995).

7. Chris Chesher writes convincingly on “The Ontology of Digital Domains” involved in this experience (Holmes 1997, 79ff.).

8. Robert Hanke uses categories proposed by Pierre Bourdieu to give an account of these developments: <http://hhobel.phl.univie.ac.at/gl/gl9506/msg00062.html>.

9. On the issue of disembodiment compare James and Carkeek (1997) as well as Wilson (1997). See also Featherstone and Barrows (1995).

10. A chronicle of events and several political assessments can be found at <http://www.univie.ac.at/philosophie/facts/sparfl/sparfl.html>.

11. Mark Dery (1994) has written lucidly on the postmodern rhetorics of Cyberspace.

12. For multi-media experiences cf. Chapter 7 in Jones (1997) and Barrett (1992).

13. <http://hhobel.phl.univie.ac.at/register.html>

14. <http://www.rzuser.uni-heidelberg.de/~dkoehler/Virtual Library/14.de.htm>

15. <http://www.thur.de/home/annette>

16. Recent contributions to this topic can be found in Holmes 1997. Cf. <http://www.lcl.cmu.edu/CAAE/Home/Forum/report.html>. See also <http://www.univie.ac.at/philosophie/bureau/democracy.htm> and my paper, “Could Democracy be a Unicorn?” in *Monist*, 1997, available on-line at <http://hhobel.phl.univie.ac.at/mii>.

References

Barrett, E., ed. 1989. *The Society of Text. Hypertext, Hypermedia, and the Social Construction of Information.* Cambridge, MA: MIT Press.

———. 1992. *Sociomedia: Multimedia, Hypermedia, and the Social Construction of Knowledge.* Cambridge, MA: MIT Press.

Barrett, E., and Redmond, M., eds. 1995. *Contextual Media: Media and Interpretation.* Cambridge, MA: MIT Press.

Baym, Nancy K. 1995. “The Emergence of Community in Computer-mediated Communication.” In *CyberSociety: Computer-mediated Communication and Community*, ed. Steven G. Jones, 138–63. Thousand Oaks, CA: Sage.

Biocca, F., and Levy, M., eds. 1995. *Communication in the Age of Virtual Reality.* Hillsdale, NJ: Lawrence Erlbaum.

Cherniak, W., C. Davis, and M. Deegan, eds. 1993. *The Politics of the Electronic Text.* Office for Humanities Communication Publications, Number 3. The Centre for English Studies.

Dery, M., ed. 1994. *Flame Wars: The Discourse of Cyberculture.* Durham, London: Duke University Press.

Featherstone, M., and R. Burrows, eds. 1995. *Cyberspace, Cyberbodies, Cyberpunk: Cultures of Technological Embodiment.* London: Sage.

Harasim, L. M., ed. 1993. *Global Networks: Computers and International Communication.* Cambridge, MA: MIT Press.

Holmes, D., ed. 1997. *Virtual Politics: Identity and Community in Cyberspace.* London: Sage.

Hrachovec, H., ed. 1997. *Monist Interactive Issue*, Volume 80.

Ihde, D. 1990. *Technology and the Lifeworld: From Garden to Earth.* Bloomington, Indianapolis: Indiana University Press.

James, Paul, and Freya Carkeek. 1997. "This Abstract Body: From Embodiment Symbolism to Techno-Disembodiment." In *Virtual Politics: Identity and Community in Cyberspace*, ed. D. Holmes, 107–24. London: Sage.

Jones, S. G., ed. 1995. *Cybersociety: Computer-Mediated Communication and Community.* London: Sage.

McLuhan, M., and McLuhan, E. 1988. *Laws of Media: The New Science.* Toronto: University of Toronto.

McLuhan, M., and B. R. Powers. 1989. *The Global Village: Transformations in World Life and Media in the 21st Century.* New York: Oxford University Press.

Mitchell, W. T. 1995. *City of Bits: Space, Place, and the Infobahn.* Cambridge, MA: MIT Press.

Nielsen, J. 1990. *Hypertext and Hypermedia.* San Diego: Academic Press.

Olson, D. R., and N. Torrance, eds. 1991. *Literacy and Orality.* Cambridge: Cambridge University Press.

Pressman, R. S. 1992. *Software Engineering: A Practitioner's Approach.* New York: McGraw-Hill.

Shields, R., ed. 1996. *Cultures of the Internet: Virtual Spaces, Real Histories, Living Bodies.* London: Sage.

Talbott, S. L. 1995. *The Future Does Not Compute: Transcending the Machines in Our Midst.* Sebastobol, CA: O'Reilly and Associates.

Tanenbaum, A. S. 1996. *Computer Networks*. Englewood Cliffs, NJ: Prentice-Hall.

Turkle, S. 1995. *Life on the Screen: Identity in the Age of the Internet*. New York: Simon and Schuster.

Wilson, Michelle. 1997. "Community in the Abstract: A Political and Ethical Dilemma?" In *Virtual Politics: Identity and Community in Cyberspace*, ed. D. Holmes, 145–62. London: Sage.

Winograd, T., ed. 1996. *Bringing Design to Software*. New York: Addison-Wesley.

Cultural Attitudes toward Technology and Communication: A Study in the "Multi-cultural" Environment of Switzerland

Lucienne Rey

Introduction: A Pragmatic Definition of "Culture" and Cultural/Political Lines in Switzerland

Given the title of the CATaC meeting, we could well ask ourselves what we actually understand by the word "culture." As we know, there are several hundred recognized definitions within ethnology alone, so our discussion could be endless. In this paper, however, I should like to use the term in a pragmatic sense, and equate "culture" with the idea of "speaking the same language." This definition is practical in that it not only allows to cite major differences—for instance between the English and French nations—but also covers more subtle variations such as vocabulary and accent or even finer differences such as those between the cultures of the upper classes and the ghettos. The central role played by language within a specific culture can also be observed in terms of attempts by totalitarian regimes to undermine a cultural minority by forbidding the use of its language. Current examples are the Berber language in Algeria and Kurdish in Turkey.

If we base our understanding of culture on language, then Switzerland offers ideal conditions for cultural comparisons, because it is, to a certain extent, a kind of "language laboratory." The Swiss State is divided into three major linguistic regions, whose inhabitants speak German, French, or Italian, respectively, plus an additional language community—that of Rhaeto-Romansch—which is spoken by a small minority.

And these empirically detectable external differences within the Swiss population—meaning the various language groups to which people belong—are quite obviously linked to their various inner concepts, value judgments or opinions. In other words, each language group is linked to a set of highly specific attitudes. This is repeatedly revealed during referendums, where the difference of opinion often clearly follows the linguistic borders.

For example, I would like to mention the voting results obtained when the Swiss voted on joining the European Economic Area. It was quite clear: in the French-speaking area, a majority voted in favour of joining, while in the German- and Italian-speaking areas, opposition to membership took the upper hand. In Swiss politics, and particularly regarding votes on foreign and environmental policy, this dichotomy occurs frequently. The French-speaking area—sometimes together with the Ticino—constantly finds itself in the role of the political loser. The French-speaking Swiss were particularly indignant about the EEA vote, since the results were extremely close—membership was rejected with a majority of 50.3%! That such dominance by the German-speaking area arouses a certain amount of resentment in its French counterpart is thus understandable. In short, empirically measurable differences in attitudes clearly correlate with the linguistically defined cultures of Switzerland.

Cultural Attitudes towards Traditional Mass Communication: Differences Among Linguistic Groups

These patterns, of course, are not obviously linked to communications technology. But I shall now come closer to the phenomenon of these new technologies by placing them in the context of other, more traditional communications media and situations. I assume that the media, including newspapers, reflect a specific cultural group and are at the same time its mouthpiece. I shall therefore attempt to develop hypotheses concerning attitudes towards the use of the Internet in the three major language areas, based on various scientific investigations in media and mass culture.

The population rarely takes a theoretical approach to the concept of "culture." On the contrary, when confronted with specific activities and situations during everyday life, people either regard these as part of culture or exclude them from it. At the end of the eighties, a large-scale research program on the "cultural identity" of the Swiss population took place. Within this framework, surveys were carried out to determine the population's concept of culture. A methodical approach with photographs was used: interviewees were shown pictures of everyday situations and asked to state whether or not these situations represented "culture" from their point of view. The "cultural" situations assessed differently by members of the various language groups are shown in Figure 1.

This figure reveals two notions of culture, as articulated by the author of the study:

> The Latin areas show a systematically stronger tendency to accept technically portrayed culture within the overall notion of the term, meaning the consumption of newspapers and magazines, television and computers. Common to all the situations was the fact that they included forms of modern mass culture and markets and took the form of indirect publicity. An intermediary was always present between organising consumption and the act itself: direct contact with a public was excluded or secondary. In the German-speaking area, reaction to this was diametrically opposite: those approached distanced themselves from this attitude. Here, situations with a better chance of

Figure 1
Cultural Elements and Linguistic Groups

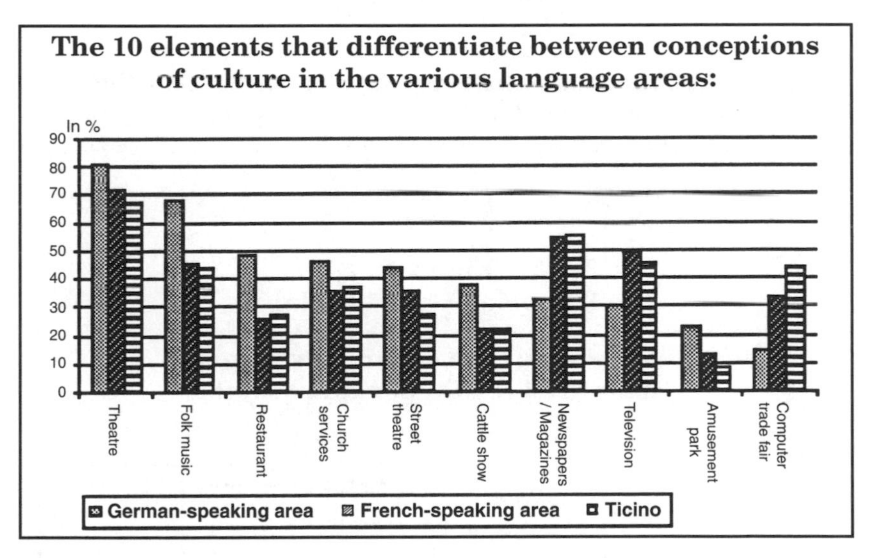

Source: Hans-Peter Meier-Dallach, 1991

> being appreciated as forms of personal culture were those with direct publicity, i.e., a specific public. Contacts within a large or small circle were included. Popular, folkloric events held in small, local premises or scenes of agricultural work were those that better characterised the conception of culture in the German-speaking area of Switzerland. (Meier-Dallach 1991, 14: translation by the author, LR).

These findings, moreover, are confirmed by statistical enquiries concerning media use: television is more used in the Latin parts of Switzerland than in the German part (Figure 2).

In other words, the Latin communities are more open to technical methods and forms of communication. This finding should also be relevant to the newest of such media, the Internet and the World Wide Web. Indeed, this empirical finding leads us to our primary working hypothesis: this finding suggests that the Internet will be more present and used to a greater extent in the Latin language areas than in the German-speaking area.

Cultural Differences in Attitudes toward Technology: "Cantons" and "Communes" as Representing the Swiss Population

Now we must find empirical evidence for this hypothesis. From the point of view of the social scientist, a large enquiry among the population of Switzerland would be the most effective approach—but admittedly also the most expensive. So, for this more modest study I

Figure 2
Television Consumption

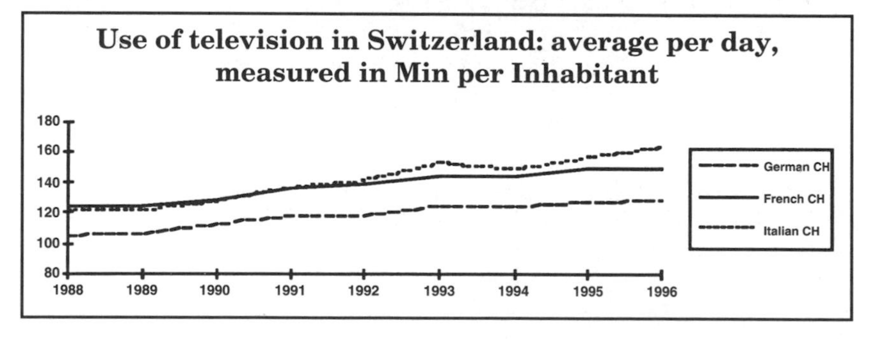

chose to analyse how various civil/political administrations in Switzerland behave on the Internet.

Switzerland today consists of twenty-six cantons. The cantons—often also referred to as "the States"—are the original States which joined together to form the Federation to which they transferred part of their sovereignty in 1848. Each canton and half-canton has its own constitution, parliament, government and courts. French is spoken in four cantons, Italian and Rhaeto-Romanch in one, respectively, and German in eighteen; two cantons, where German and French is spoken, are thus bilingual (see Figure 3).

In 1997, the Swiss "Institut de hautes études en administration publique" (IDHEAP) in Lausanne carried out a study on "cyber-administration" in Switzerland. The authors found that at the beginning of 1997, 73% of the cantons were present on the Internet. The only cantons which had no intention of becoming active on the Net were located in the German-speaking part of Switzerland (Poupa et al. 1997, 17). At least with regard to the development of

Figure 3
The Language Areas in Switzerland

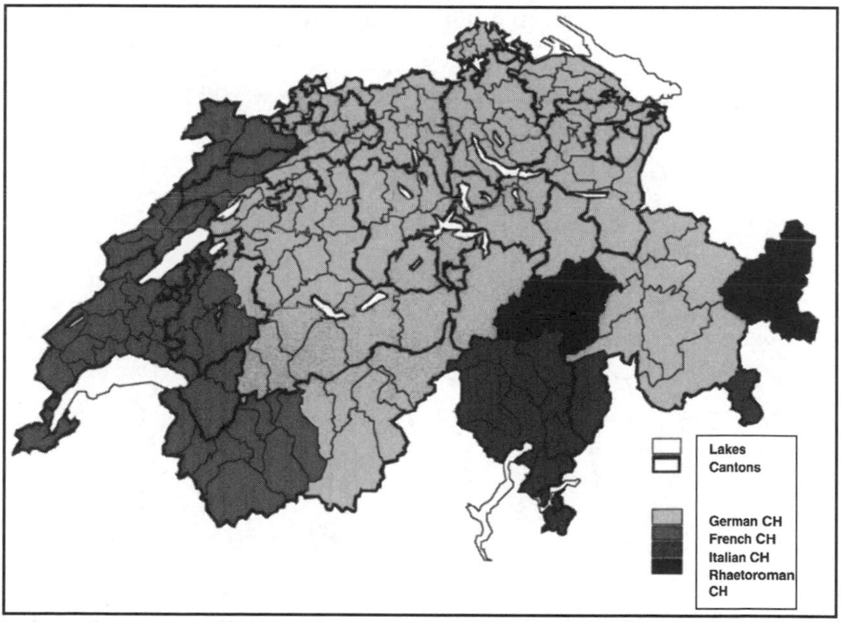

Source: Basemap © SFSO Geostat / S+F

"cyberadministration" on the cantonal level, the German part of Switzerland shows a certain delay in developing Internet activities, even if today (in July 1998) this retardation is made up.

At the lowest level, the Swiss Confederation consists of local political authorities—the communes—which, according to the Swiss Federal Chancellery (1998: 3), currently number 2,942. This number is diminishing as local authorities combine. This is shown also by the indications in the Swiss yearbook of the public life, *Publicus*, which specify the number of communes with 3,015. As the indications in *Publicus* are more detailed, I will base my work on this data.

In addition to the tasks entrusted to them by the Federation (such as registering the population and civil defense), the local authorities also have specific responsibilities for education and social welfare, energy supply, road building, local planning, taxation, etc. To a large extent, these powers are self-regulated. Still according to the *Publicus*, there are a total of 3,015 communes in Switzerland: 1,768 are German-speaking, 905 are French-speaking, 270 are in the Italian area and, finally, Rhaeto-Romansch is spoken in 72 (Schwabe and Co. 1997, 153).

Figure 4
Homepages of Swiss Communes with more than 5,000 Inhabitants

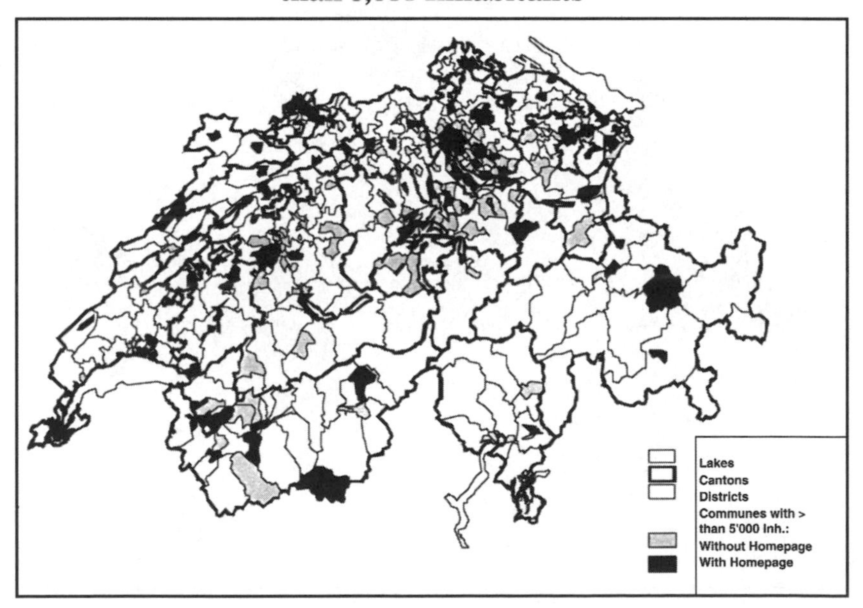

Source: Basemap © SFSO Geostat / S+F

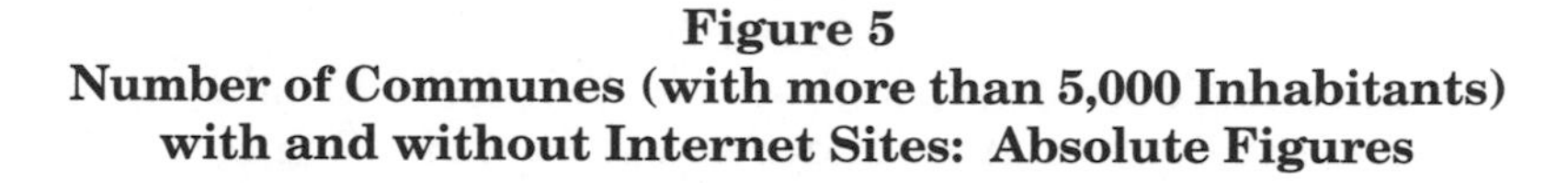

Figure 5
Number of Communes (with more than 5,000 Inhabitants) with and without Internet Sites: Absolute Figures

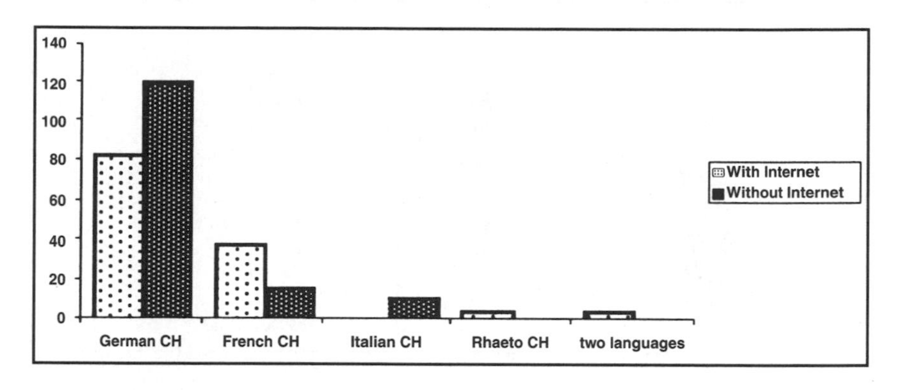

For this study, I examined whether each commune of over five thousand inhabitants had an Internet site. My approach was to type in the name of the commune as a domain. When the Internet announced that the corresponding "host" could not be found, I checked via a search engine (Alta Vista) to make sure.

The results of my research are shown in Figure 4. The first thing that strikes us is that the majority of the communes with over 5,000 inhabitants are found around the large urban areas, most of which are in the middle of the country. The larger agricultural regions in the foothills of the Alps, the Alps and the Jura mountains, as well as the agrarian hinterlands of the Canton of Vaud remain white.

If we concentrate on the relationship between the light and dark grey areas, it is clear that in the French-speaking area, the dark grey areas (i.e., communes with homepages) predominate; by contrast, there is a greater proportion of light grey (i.e., communes without homepages) in the German-speaking area. The canton of Geneva is a special case: this canton has obviously coordinated the Internet sites of the various communes, so even the smaller communes in the canton have their own Internet site, and all with a uniform style. Basel's site is also different, since its homepage is specifically stated as a cantonal one: it cannot be found under the name of the city, Basel, but under the cantonal abbreviations of BS or BL. Basel-city was indeed the first Swiss canton with a homepage, which was created in December 1994 (Poupa et al. 1997, 16). Finally, Ticino is especially underrepresented, since only

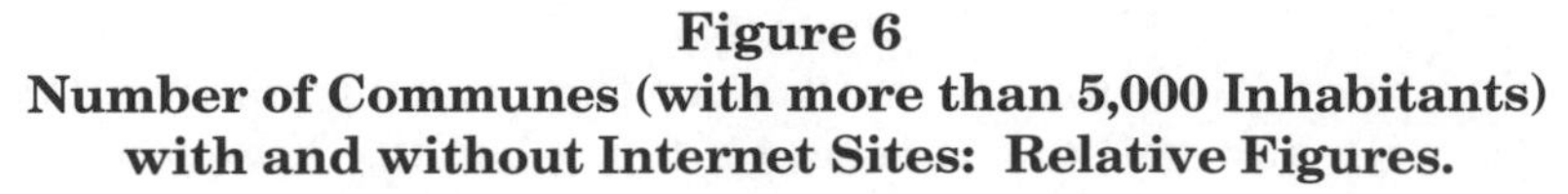

Figure 6
Number of Communes (with more than 5,000 Inhabitants) with and without Internet Sites: Relative Figures.

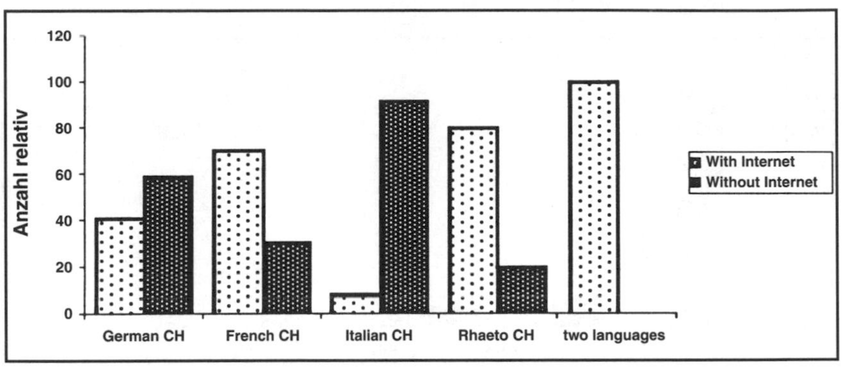

Bellinzona can be found in the "virtual world." This area does not come up to expectations, but the reason for this could be that the canton as a whole is highly oriented towards tourism. Its tourist association is correspondingly active on the Internet, and could possibly render efforts on the part of the individual communes to join the Web superfluous. It is not completely surprising, however, that Ticino's approach is different to that of the French area of Switzerland: when voting, it is frequently evident that the Italian language area's affinities align themselves with the opinions of the "German camp" at times and with those of the French area at others.

The tables shown in Figures 5 and 6 make the predominance of Latin Switzerland on the Web even clearer; relatively speaking, in the French- and Rhaeto-Romansch-speaking area there are many more communes with over five thousand inhabitants present on the Web than in the German part of the country or Ticino.

Conclusions

Based on this (reasonably reliable) data, it is possible to confirm (or in Popper's sense, it is not possible to reject) the initial hypothesis that use of the Internet and the Web is indeed more widespread in the Latin-speaking area of Switzerland, at least as this is represented by the activities of the communal authorities.

However, this eloquent fact as yet clarifies nothing. And here, we are confronted with the limits of empirical social research and move towards philosophy. There are philosophers and historians who believe that German scepticism towards progress has a historical basis. The Romantic period in particular led to an enlightened approach to nature and a rejection of science and technology (Abrams 1971, 181; Heiland 1992, 47–61; Sieferle 1984, 30–56). This could be one explanation.

I also support one of Carleen Maitland's propositions (1999). According to her Proposition Five, cultures marked by low ethnocentrism will begin diffusion of interactive networks before cultures marked by high ethnocentrism. This proposition fits with the results for the Swiss referendum on membership in the European Economic Area (mentioned above), which shows that the French part of Switzerland is more open to international collaboration. In the terms of Maitland's Proposition Five, French-speaking Switzerland is thus far less ethnocentric than its German-speaking counterpart. In this connection, it is also interesting to see that Basel—the first Swiss commune with an Internet site—voted in favor of membership in the European Community.

Perhaps, too, these differences derive from something much more commonplace: it is possible that the inclination to play is more widespread in the French-speaking area, and that this takes the form of a less inhibited approach to the new media and technologies. Seen in this way, the results reflect a certain French "lightness of being."

Note

The exciting and inspiring CATaC conference would not have occurred without the initiative of Charles Ess and Fay Sudweeks. I want to thank both of them for their great efforts, as these led to both a humanly enriching exchange of ideas and empirical progress. I also want to thank them for their help with developing the English version of this text. In addition, I want to warmly thank Hans-Ulrich Zaugg at the Bundesamt für Statistik [Federal Office for Statistics], who developed the cartographic representations of my analyses. Finally, I also thank Hans-Peter Meier (*cultur prospectiv*), who let me draw generously from his wealth of experience and sociological findings.

This manuscript appeared originally in the *Electronic Journal of Communication/La revue electronique de communication*, 8 (3 & 4), 1998 (see <http://www.cios.org/www/ejcrec2.htm>) and is reprinted by kind permission of the editors.

References

Abrams, Meyer Howard. 1971. *Natural Supernaturalism: Tradition and Revolution in Romantic Literature*. New York: W.W. Norton and Company.

Bundesamt für Statistik. 1997. *Statistisches Jahrbuch der Schweiz 1998*. Zürich: Verlag Neue Zürcher Zeitung.

———. 1994. *Statistisches Jahrbuch der Schweiz 1994*. Zürich: Verlag Neue Zürcher Zeitung.

Heiland, Stefan. 1992. *Naturverständnis. Dimensionen des menschlichen Naturbezugs*. Darmstadt: Wissenschaftliche Buchgesellschaft.

Maitland, Carleen. 1999. "Global Diffusion of Interactive Networks: The Impact of Culture." *AI and Society* 13: 341–56.

Meier-Dallach, Hans-Peter. 1991. *Das Kulturverhalten der Bevölkerung: Vielfalt, Kontraste und Gemeinsamkeiten*. Bern: Schweizerischer Nationalfonds zur Förderung der Wissenschaftlichen Forschung: Nationales Forschungsprogramm 21: Kulturelle Vielfalt und nationale Identität. Reihe: Kurzfassungen der Projekte.

Poupa, Christine, et al. 1997. *La cyberadministration en Suisse*. Lausanne: Institut de hautes études en administration publique (IDHEAP). Discussion paper no 6.

Schwabe and Co. AG. 1997. *Publicus 1997/98*. Schweizer Jahrbuch des öffentlichen Lebens / Annuaire suisse de la vie publique. Basel: Schwabe and Co. AG.

Sieferle, Rolf Peter. 1984. *Fortschrittsfeinde? Oposition gegen Technik und Industrie von der Romantik bis zur Gegenwart*. München: C.H. Beck.

Swiss Federal Chancellery. 1998. *The Swiss Confederation: A Brief Guide*. Berne: Swiss Federal Chancellery.

Diversity in On-Line Discussions: A Study of Cultural and Gender Differences in Listservs

Concetta M. Stewart, Stella F. Shields, Nandini Sen

Introduction

The emergence of a global information infrastructure (GII) has created more opportunities for multicultural communication in the form of "on-line communities." By their nature these communities are as diverse as the technologies of the GII. A commonly-held belief is that these on-line communities are also naturally democratic and open. As a result, important issues are being raised concerning how these different groups are coping with new technologies and what role factors such as gender and culture play in participation in and creation of these new systems.

The Internet draws together people of different cultural groups, both locally and around the world. There is a problem, however, with the predominant models of policy and research that assume that democratic participation in networked systems simply comes from the equal availability of technology and the necessary skills to access these systems. Previous research has shown that this is not the case (Balka 1993; Ebben and Kramerae 1993; Herring 1996; Herring 1993; Kramerae and Taylor 1993). Instead, we have seen that even with the most basic of systems, women and minorities are not participating in anywhere near equal numbers (Nielson Media Research 1996; Spender 1995; Stewart et al. 1997). Consequently, a key goal of research in this area must be to inform the creation of policies to improve the equity of access and use of these new technologies by all groups.

The primary focus of this research, then, is how groups who are typically absent from these on-line communities, such as women and minorities, can participate more equally. This lack of participation in the GII has implications for the economic and social well-being of those excluded as well as for the larger global community. This work, therefore, also has implications at a global level as we look at the lack of participation of developing nations in the evolution of a global information infrastructure.

Cross-Cultural Communication

Cross-cultural communication can be defined as consisting of intercultural, multi-domestic, and cross-gender communication or genderlect communication (Tannen 1990). Key issues in cross-cultural communication research include styles of conflict and negotiation (Ting-Toomey 1985) and construction of identity and self-disclosure (Ting-Toomey 1988) in interpersonal and group contexts. In cross-cultural communication, meaning and interpretations are derived both collectively and individually through interaction: collectively, in the sense that meanings are negotiated between communicators and, individually, because the process of interaction is mediated by individual perceptions that are subject to one's identity and expectations which are in turn guided by culture (Gudykunst and Kim 1996). Thus, it may be argued that the culture in which norms are developed will be reflected in all interactions regardless of the communication medium. It has been widely recognized in cross-cultural research that people derive different meaning and often key information, however, from the contextual aspects of the interaction (Hall 1976). Consequently, it is critical to determine how such cultural norms affect communication processes in the context of mediated communication. Unfortunately, though, while there are significant bodies of research on both intercultural and mediated communication, cross-cultural communication via electronic media has largely been overlooked (Ma 1996).

High- and Low-Context Cultures and Communication

Based on his extensive study of cultures around the world, Hofstede (1983) identified four common dimensions upon which cultures could be compared: power distance or the extent to which less powerful members of society accept that power is distributed unequally; masculinity or when there are clearly defined sex roles with male values of success, money and possessions as dominant; uncertainty avoidance or the extent to which people feel threatened by ambiguity; and individualism which reflects the relational ties between an individual and others. Although this research focused on cultures of different nations, it can be argued that Hofstede's findings can also be applied to a variety of cross-cultural communication situations.

Scholars of cross-cultural communication, most notably Hall (1976) and Ting-Toomey (1988), regard Hofstede's dimension of individualism as a crucial dimension of variability across cultures. It is also a key dimension in understanding interpersonal and group interaction and communication processes. In an individualistic culture, individuals are loosely integrated with others and value their own self-interest and that of their immediate family only. In contrast, in collectivistic cultures, individuals relate to larger collectivities and groupings and themselves as integrated with the whole.

Hall (1976) describes cultures as being high- or low-context, with context serving as the information that surrounds and gives meaning to an event. In other words, in high-context cultures, meaning is found in the nature of the situation and relationships, while in low-context cultures meaning is found in the words. Furthermore, key to interpersonal and communication behavior, high-context cultures strive for subtlety, patience and empathy, while low-context cultures value straight talk, assertiveness and honesty. Hall explains that high-context cultures also value collective needs and goals and create "us-them" categories, while low-context cultures value individual needs and goals and believe that every individual is unique.

Ting-Toomey (1988) has developed Face-Negotiation Theory to explain cultural differences in a key communication context, negotiation and conflict. Her basic assumption is that all people negotiate face, with face serving as a metaphor for public self-image. Face Work involves enactment of face strategies, verbal and non-verbal moves, self-presentation acts, and impression management interaction. Our identity can always be called into question and this leads to conflict and vulnerability; however, this varies from culture to culture, particularly along the dimension of high- and low-context cultures. Ting-Toomey (1988) describes this issue of identity and vulnerability in terms of the "faces of face." For example, in high-context cultures one strives to preserve the other's autonomy through face-saving and to include the other through face-giving, while in low-context cultures, one seeks to preserve one's own autonomy through face-restoration and to include oneself through face-assertion. In conflict resolution and negotiation, communication styles vary based on concern for self- and other-face. Her research also suggests that there is in fact a strong relationship between culture and face concern in conflict resolution and negotiation, with high-context favoring other-face and avoiding, obliging, compromising, and integrating, and low-context favoring self-face and dominating.

Gender and Communication

Tannen (1990) believes that gender differences can also best be observed from a cross-cultural approach, one that does not assume that differences arise from men's efforts to dominate women. The belief that masculine and feminine styles of discourse are best viewed as two distinct cultural dialects, rather than as inferior or superior ways of speaking, typifies this stance and is summed up by the term "genderlect." While some scholars do not believe that identifying gendered communication styles is important or even appropriate, Herring (1996) and Tannen (1990) believe that ignoring those differences creates a greater risk than does the danger of naming them.

A significant body of research on the fundamental issue of gender differences and communication practices exists (Lakoff 1973; Rakow 1986; Spender 1985; Stewart and Ting-Toomey 1987; Tannen 1994). However, as Rakow (1986) states, we need to refocus this research away from a conceptualization of gender as an individual attribute to bring more attention to the structures of the relationship between gender and power. The extent of the problem is dramatically illustrated by research which finds that men perceive women as dominating a discussion even when they contribute as little as 30% of the talk (Herring, Johnson, and DiBenedetto 1992; Spender 1989). Spender (1989) explains this finding by observing that since it is the "natural order of things" for women to contribute significantly less to a group discussion than their male counterparts, women are then thought of as dominating the discussion when they participate at anywhere beyond that minimal level.

A common perception, however, is that women talk more than men. Tannen (1993) states that the context is essential to explaining this misconception. For instance, research has shown that men talk more in formal versus informal tasks and more in public versus private communication. The effect is that while same-sex task teams produce consistent amount of output, in mixed sex teams, the men produce more than the women (James and Drakich 1993; Rakow 1988). In public spaces, for instance, men speak for a greater length of time and men's speech is more on task while women's is more reinforcing. Men's talk serves to hold floor for extended lengths of time, so that talking exercises dominance and prevents others from speaking. The ultimate effect is a lack of regard for women and their speech. This dominance also implies higher social status and that men are more competent to complete the tasks or to discuss the issues at hand than are women (James and Drakich 1993). Tannen (1994) explains that women also typically use more supportive language patterns, which thereby diminishes the power of their own contributions. There are obvious implications for women, then, as they are increasingly participating in public arenas such as the workplace and politics where they may not have equal opportunity for participation.

Women also strive less actively for control (Nadler and Nader 1987), whereas societal expectations are that men will dominate task-oriented discussions. This conversation dominance is evidenced by amount of communication and amount of interruptions, with male dominance in speaking time achieved through interruptions (James and Clark 1993). Lakoff (1995) believes that this control of the discussion is interpretive control and goes beyond the genderlect (i.e., simply a difference in language style based on gender) Tannen describes. According to Lakoff, men are actually assigning valuation to women's speech. She also contends that men will also use silence, since to ignore is also a sign of power; non-response is one of the most effective ways the powerful silence the less powerful. She states that as "annoying and discouraging as interruption is . . . non-response is by contrast annihilating" (1995, 28) because to ignore someone is to deny their existence.

Tannen (1993) states that while scholars recognize intuitively that interruption and topic control in conversation is encouraged by, and encourages, power imbalance, research has shown that women actually interrupt more. She admits that this finding was puzzling until still other research showed that there was, in fact, a difference in the patterns of interruption. For instance, men raise more new topics than women and use interruptions to change subject and take floor, while women use interruptions as cooperative overlap and to show support for the speaker. Tannen (1993) also identifies another key difference in the communication practices of men and women, i.e., men use a more adversarial style in discussions, while women are likely to ask more questions. Women also use verb qualifiers and have a pattern of politeness behaviors, leading to an image of less intelligence. According to Lakoff (1995), women have learned the language of apology, and these linguistic patterns negatively affect credibility and suggests uncertainty and triviality in the subject matter.

Herring's (1996b) research on Internet listserv discussions supports these differences in communication patterns and has shown that men are more critical, flaming and adversarial. Men also value freedom from censorship along with candor and debate and will violate negative politeness (i.e., imposition) with the longest posts, copying most text and the longest signature files. Women value harmony and will avoid conflict, controlling action to minimize damage, which is a positive politeness pattern.

Typically it is the most dominant and powerful group whose values take on a normative status. Herring (1996a) contends that the issue then is to understand whose values inform the rules of behavior on the Internet. These differences that reproduce patterns of dominance must be known and understood in order that we may address them to achieve a more equitable and hospitable environment in cyberspace.

Research Questions

The focus of this research is related to how differences between the communication styles of the different cultures (high- and low-context, male and female) exhibit themselves in this mediated communication environment. The primary research questions, then, are:

RQ1: Is there a predominant cultural style?

What was the message frequency by individual and by group based on culture? Based on gender?

What was the message length by individual and by group based on culture? Based on gender?

What was the adoption rate by individual and by group based on culture? Based on gender?

RQ2: Are there differences in communication styles of men and women?

Is there evidence of collective versus individual concerns?

Is there evidence of self- versus other-face orientation

RQ3: Are there differences in communication styles of white Americans and the other cultures?

Is there evidence of collective versus individual concerns?

Is there evidence of self- versus other-face orientation?

The Study

This research goes beyond the question of availability and technical proficiency to examine cultural and gender differences in communication patterns, and how these differences specifically affect who actually controls and directs these on-line discussions.

Listservs are only one form of on-line communication and are used for maintaining e-mail-based distribution lists on the Internet. Anyone who can send and receive Internet e-mail can access a listserv depending on its owner's permission. To subscribe to the listserv a potential user needs only to send an e-mail message to the listservs system. The listserv studied here was set up for graduate and undergraduate students enrolled in a global telecommunications course at a major urban university in the United States. Other purposes of this implementation were also to explore the feasibility of listservs for enhancing the live classroom experience—with the possibility of using this technology as one component of a distance learning environment—as well as to evaluate more closely the characteristics and effectiveness of the group discussion process using this technology.

This listserv was intended to create an open dialogue on topics related to the class. Given previous research (Balka 1993; Herring 1993), however, showing the tendency of a small group of individuals to dominate listserv discussions, the instructor established "netiquette" or guidelines for communication behavior in the discussion. These guidelines included no flaming (or personal attacks), no shouting (the use of all capital letters), no personal messages, and no really long messages.

Topics were generally raised by the students and were related to the course material in global telecommunications. In addition, the graduate students were given the further responsibility of keeping meaningful discussion going on the listserv. The topics addressed the impact of technology around the world from social as well as political and economic perspectives, and issues of cultural diversity and gender featured prominently in many of the discussions.

Twenty-two people, consisting of graduate and upper-division undergraduate students, the course instructor and a guest instructor, participated in the listserv discussion for four months. The participants represented a broad range of ages from twenty to fifty, though most were twenty to thirty years of age. The class met in-person once per week and participated in the listserv discussion throughout the week. There were nine males and thirteen females on the listserv, consisting of six Asians, one African, six African Americans, one Latin American, and eight white Americans. Therefore, given its diversity, this was thought to be an especially appropriate group for such a study as posed here.

Analysis

A complete transcription of the listserv discussions was collected for one semester, or fifteen weeks, and consisted of over three hundred[1] messages. Included in these transcripts were message header information such as sender, date and time, and subject[2], as well as the actual content of the messages. Multiple methods were used in the analysis of these data. Patterns of interaction were studied by examining ratios of message frequency, message length, and rates of adoption as well as language used and topics raised. A close reading of the transcripts was conducted based on the notions of individualism-collectivism in cultures (Hofstede 1983), high- and low-context cultures (Hall 1976), "face work" (Ting-Toomey 1988) and genderlect (Tannen 1990). These theories informed the analysis of the listserv discussion as well as the examination of other cultural and gender differences in the on-line communication process. For cultural comparisons, one group consisting of white Americans was categorized as individualistic, while the other group consisting of African American, Latin American, Asian and African individuals was categorized collectivistic. (For more details on this categorization scheme, see Hofstede 1983.)

Results

Below are presented the results of the analyses including: message frequency, message length, adoption rates and conversational analysis.

Message Frequency

The analysis shown in Table 1 supports finding of previous studies, i.e., men sent more than twice as many messages in total as women, with the men sending 204 messages as compared to the women who sent 100 messages. The difference in volume here is more striking when one considers that there were eleven women in the group and nine men. The average number of messages per person by gender perhaps makes this point more clearly, revealing 22.7 messages per male and 9.0 per female. In addition, as also shown in Table 1, white Americans including both men and women sent more messages than men and women of other cultural groups combined, with white Americans sending 159 messages versus 145 messages sent by the others, and despite the fact that there were only six white Americans out of the total group of 20. The cultural groups represented in this other category includes six African Americans, one Latin American, six Asians[3], and one African. When the average number of messages is calculated by cultural grouping, a similar pattern emerges. White Americans sent more than twice as many messages, 26.5 messages on average, while the others sent 10.3 messages per person.

Table 1
Number of Messages by Sender by Gender and by Culture

	Number of Messages			
Sender	**Male**	**Female**	**White American**	**Other**
M1	80		80	
F1		30	30	
M2	24			24
M3	24			24
M4	23		23	
M5	19			19
F2		19		19
M6	13			13
F3		13		12
M7	13		13	
F4		13		13
F5		12	12	
M8	7			7
F6		6		6
F7		5		5
F8		1		1
F9		1		1
M9	1		1	
F10		0		0
F11		0		0
Subtotals	204	100	159	145
TOTAL		304		304

Message Length

When examined by the length of message[4] some interesting dynamics are observed as well (table 2). While there were equal numbers of men and women sending long messages, the men sent more long messages in total than did the women (fifty-one messages versus thirty-four messages), or an average of 7.3 messages for males versus 4.9 for females. When looking at this behavior according to culture, the averages were similar, with white Americans sending 7.4 long messages on average and the others sending an average of 5.3 long messages. It is worth noting here that it was an Asian woman who sent the longest original messages, consisting of both complex and thought-provoking discussion, while it was an Asian male who sent the most long messages that consisted of replies of a few sentences with the entire original message copied in most every instance.

Rates of Adoption[5]

In Figure 1, we see that males adopted the technology first, most of them doing so in the first two weeks, while more than half of the fe-

Table 2
Number of Long Messages by Sender

	Number of Long Messages			
Sender	**Male**	**Female**	**White American**	**Other**
M3	14			14
M1	12		12	
F2		9		9
M4	8		8	
F5		7	7	
M5	6			6
F3		5		5
M2	5			5
F1		5	5	
M7	5		5	
F6		4		4
F4		2		2
F7		2		2
M6	1			1
Totals	51	34	37	48

males adopted in Week 6 or later. It is also worth noting that the only two non-adopters (shown as Week 11) were women. When viewed by culture, in Figure 2, we see that all but one white American adopted in the first two weeks, while a majority of the members of the other groups adopted after Week 3. Again, it is worth noting that the only two non-adopters (shown as Week 11) were in this category as well.

Conversational Dynamics

There were some notable differences in communication styles with respect to cultural and gender differences revealed in the analyses of the transcripts. One key area of difference seems to center on how differences of opinion are handled and whether or not there is a perception of "winning" or "losing" in the process. Drawing from Lakoff (1975, 1979, 1990), Tannen (1990) explains that systematic differences in conversational style can lead to misunderstandings in both cross-cultural and cross-gender communication. Citing Gumperz (1982), Tannen (1990) also describes the best method to discover what is going on is to look for key episodes where communication has broken down. This process involves "identifying segments in which trouble is evident" and "looking for culturally patterned differences in signaling meaning that could account for trouble" (6).

Figure 1
Adoption by Week by Gender

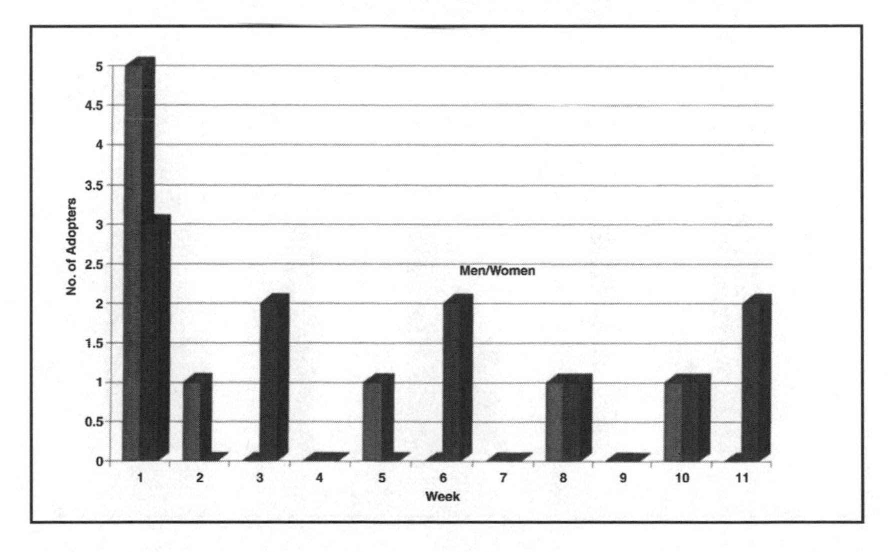

Some of these key segments will be presented below.

Segment 1

This is an example of one of the more heated exchanges.

F2, an Asian female observes:

> What is considered censorship in one country is thought of as protection in another.

M1, a white American male responds:

> Are other cultures so far gone that people have mindsets that they have no control over? I am a person first, a US citizen second. Responsibility is not a cultural issue. The application of responsibility is manifest in individuals based on learned cultural impressions. The Internet offers new thinking, new mindsets. Perhaps (in the future) the Internet will cause the downfall of current geo-political authority (I hope). Culture

Figure 2
Adoption by Week by Culture

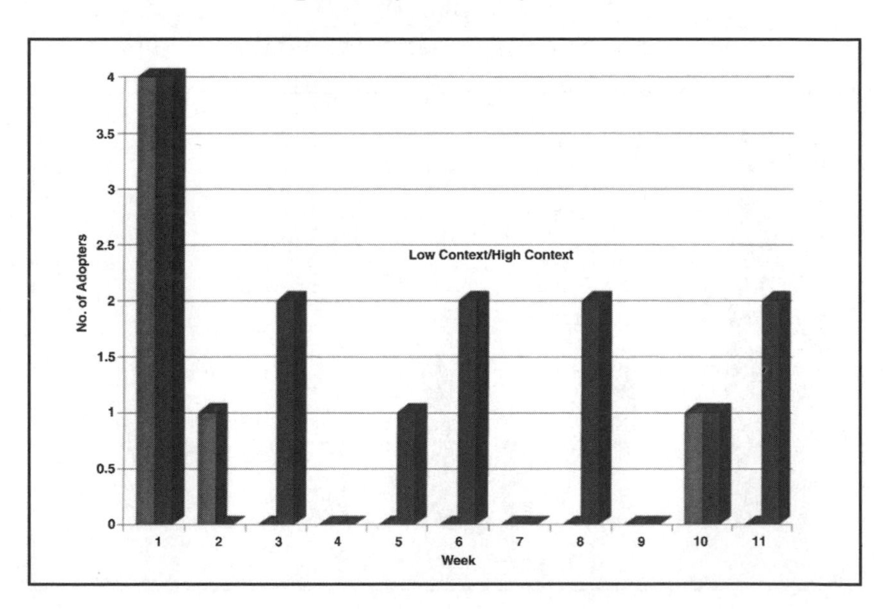

> makes me sick and for an entire people to be dog-leashed by tradition and parental/government/religious control is equally sickening. The Internet is mental anarchy. No authority has the right to impose on my thinking—nor anyone elses.
>
> What I'm trying to say is you as an individual should have the right, no matter where your heritage or national boundries lie, to choose your own heaven or hell.

F2 replies:

> M1, first of all, stay off the emotions and stick to a decent, polite academic discussion . . . This is not a conflict that needs to be resolved. So I'll end the discussion here.

M1 responds:

> F2, My opinions are quite polite, extreme, I admit but tactfully conveyed.

M5, a Latin American male, comments:

> Hey guys, . . . Culture is an extremely complex issue . . . different opinions will always reign . . .

M1 closes the discussion with:

> M5 . . . I ask that the class agree that there is good and bad in all cultures.

In this exchange, M1 places a strong emphasis on individualism and self-determination, while the others are comfortable with that cultures will differ in this respect. M1 also seems to welcome conflict and needs a final, clear-cut resolution, with everyone in agreement. He uses a dominating style as described by Ting-Toomey (1991), while M5 is seeking compromise, and F2 is avoiding conflict which is also consistent with Ting-Toomey's theoretical framework.

Segment 2

In this conversation, the instructor posts a message from "New Thinking," a free weekly e-mail "contributing to a philosophy for The Digital Age," by Gerry McGovern, in which McGovern states that:

"Freedom, privacy and censorship are linked. To give a certain level of freedom and privacy to one person, one must inevitably censor/regulate the privacy and freedom of another."

F3, an African-American female:

> In regards to McGovern's essay, I don't think it's new thinking, I think it's old thinking. All societies have regulations, and histories of it, therefore I find the author's argument for a regulated cyber-society in contrast to the theory that the on-line world is where many people might seek refuge in a constrictive environment. Those who can, are getting an understanding of the limits and boundaries of a interactive world, today, but in the future I feel that each nation should define their involvement in a global infrastructure according to their cultural value of space. By setting their global clocks by space, it will allow each nation identity to develop a threshold for which it sets the parameters. While it is the responsibility of the information rich to set guidelines for the inclusion of the information poor, it is also the manifestation of spatial identity on the part of the underdeveloped and underrepresented. It is the test of the leaders of the developing nations as to how well they conserve the interest of their countries. It's time to play the economic hand we've all been dealt. The highest ideal we should be striving for is a sense of unison, not a lineal set of regulations.

M1, a white American male, responds:

> F3, You sound embittered. . . . You have some really good points and a good feel for "Blarney" detection.

F3 responds:

> Uh M1, would you mind e-mailing me personally and telling me where I sounded "embittered?" From what I interpreted from your analysis, you made the same points I did. And, who is "Blarney" to you? Let's stick with passing judgements on global telecommunications, not on each other. I am not impressed or amused by your psychoanalytic interpretations.

M1 was perhaps applying "male bonding" rules and trying to kid around, while F3 saw the comment as highly personal and insulting. M1's choice of words would also suggest a lack of concern for other-face.

Segment 3

In another episode, an Asian male (M2) offers some advice and information to the class.

M2's message:

> Subject: Helpful hints for this week's work
>
> Hi everyone.
>
> Here are the list of journals that you will find at <our university's> library and at <another university's library>. I strongly urge everyone in the class to use <the other university's library> instead of <our> library. (It is a fact that <our libary> does not have many resources . . .)

Replies follow from several different individuals.

M1, a white American male:

> For me, it's downright impractical.

M7, a white American male:

> Due to my busy schedule, <our library> will be my only resource . . .

F1, white American female:

> I agree that <the other library> is an option.

M4, white American male:

> Little things like getting mugged, robbed, shot, stabbed, etc. prevent me from being as resourceful as others.

None of the respondents ever directed their complaints and criticisms at M2, and instead address their rebuttals to "others" or "some people." These comments also suggest an orientation toward the self rather than the group. The one exception was F1, a white American female, whose remark was in agreement with M2's original suggestion. In addition, while the men's comments might appear as openness and self-disclosure, which Ting-Toomey (1985) characterizes as leading to vulnerability, a closer reading reveals that these messages are not intimate and that the senders may actually be outlining their boundaries to the rest of the group. Another interpretation may also be that these individuals are self-confident enough to reveal this information about themselves and don't feel vulnerable at all.

Segment 4

Another interesting exchange involved M7's (a white American male) banner.

M7:

> *************************In the military, you can be a REAL man!****

M1, a white American male:

> What's this with the military quote? I spent four years as a U.S. army paratrooper.

M4, a white American male:

> . . . since I'm a psuedo-man (I'm not in the army so I can't be real) . . .

The replies were both directly and indirectly critical of M7's banner, and not the actual topic of the message. These statements could also be illustrative of a lack of concern for other's (M7's) face. They seem more concerned with self-face as their statements were about who they were, rather than who M7 was or what he might have actually meant by his banner. Since this conversation occurred very early in the listserv, it serves as a good example of problems that can arise from a lack of context. In other words, these other students did not know M7 yet, and as a result were immediately threatened or even offended by his banner without any of the context that would typically arise in face-to-face interactions. That discussion did not continue, though, as M7 immediately withdrew his banner.

Collective versus Individualistic Interaction

Some of the dialogue also provides opportunities to examine collective versus individualistic perspectives on the class activities and issues.

Example 1

Early in the list there was a discussion about the amount of work required by the course, particularly the responsibility to participate in the listserv regularly. This had not been expected of the students in other courses they had taken at the institution, so it was interpreted by some as extra work being required of them.

M1, white American male:

> I work full time and a lot of the evening courses go a little easier on students because the profs themselves are full time workers at other jobs and know the stress of family, job, school.
>
> This class is obviously different and a bit more inconveinent. I do the best I can and try to help out whoever I can.

M4, a white American Male:

> I know that every one has probably been thinking, "Why I haven't heard anything from that strikingly hansome, long-haired, sideburned, mustached, hillarious guy, with great fashion sence, on the listserv?" Well there is a good reason for that....
>
> The Top 5 reasons why M4 can't Participate:
>
> 1) Reports from other classes
>
> 2) Work

> 3) I can never get to a computer, <the satellite campus> is always full or down, and at Main, they now issue pagers, so they can page you when a computer is ready (gotta love technology!)
>
> 4) Can't access E-mail from home.
>
> 5) My unicorn had to be rushed to the vet, I was making crop circles with some friends from another solar system, I am personally arranging Elvis' comeback tour, Jimmy Hoffa, The original Paul Mc Cartney and Bruce Lee invited me to their island paradise for the weekend, and I have been taking care of abandoned Bigfoot young.
>
> I have a lot of comments about the listserv disscussion, I hope to log on again before class, I have a graduation appointment now.

M5, a Latin American male:

> I want to apologize for not being part of the discussion this week due to an outrageous schedule . . . next week you all can count on me for extra discussion and involvement.
>
> We are together in this one, and we will make it.

It was the males who offered explanations of their participation (or lack thereof), but there was a difference in the perspectives of the white Americans males versus the Latin American. M5 indicates a sense of responsibility to the collective to participate and provide support. M1's and M4's comments reflect a self-oriented perspective on the class and its requirements.

Example 2

Later in the list, the class makes observations about the group's participation in the listserv along with other group activities.

F2, an Asian female:

> It's good to hear from so many different voices this week on the listserv.
>
> F1, great job of answering the first question.

M4, a white American male:

> I have decided to let everyone know my thoughts on the class and report.
>
> You can reply, but this marks my last entry to the listserv . . .

F1, white American female:

> Perhaps we should send <the guest lecturer> an e-mail on behalf of the class, thanking him.

M2, an Asian male:

> Hi, everyone.
>
> I mentioned to <the professor> that during the holiday, we won't be able to send any messeages since <the university's> computers are going to be upgraded. Thus, she granted us to hand in the report until next Tuesday.
>
> So plan your schedule accordingly.
>
> Have a great Thanksgiving everyone!

Here the Asians and female students are offering information and help to the rest of the class, and present a collective perspective on the group's activities.

Example 3

F4, an Asian female, shares a news item:

> I notice that the list has been awfully quiet. Today I came across a news item on an on-line version of an Indian business daily, which reported that since the government had lowered import duties on foreign made information technology products, including finished goods, the Indian hardware manufacturers are likely to turn into mere vendors of products made by the multinationals. Before the government lowered the tariffs, the difference in price between an IBM PC, for instance, and its Indian counterpart, was 25%. Now, it is only 7%. The small price difference, manufacturers feel,

> will not be a disincentive against buying foreign brands. Would anybody like to comment on the implications, or report relevant news items from other parts of the world?

M5, an Asian male, replies:

> As I have seen and witnessed the trade in India. The idea here is that it is imminent that the economy will become that of a global engine. It, in many respects already is. Multi-national corporations are utilizing offshore facilities to create a Virtual Financial Sovereignty; this directly affecting governmental and social concerns. The world, if it is to fulfill the cycles of progression must succumb to the economy—as a global animal. To do business in the world today means dealing with the monster Capitalism. Indian Products are not the superior and, definitely not the less expensive. Import Tariffs exceed 300%. That means that if you buy a television here for 300.00 (U.S.$), it would cost you 900.00 (U.S.) to take it into India. This is done to encourage the purchase of domestic products. BAH HUMBUG.
>
> I'll buy the better of the two products. Thank you.

There are no replies to M5's message nor are there further comments on F4's original message. In addition, F4's participation drops sharply for the remainder of the listserv discussion.

Discussion and Implications

Some striking differences in communication patterns were observed on this listserv—by gender and by culture and perhaps there was also an interaction between culture and gender. The magnitude of these differences is particularly noteworthy, since they occurred despite the instructors' efforts to create an open, free-flowing communication environment. As outlined previously, these differences in communication patterns can be interpreted based on the systematic linguistic differences attributable to gender and culture. In other words, these patterns of communication on the listserv could be said to simply be replicating patterns of interaction that are seen in traditional face-to-face situations. These patterns do not simply mirror traditional communication environments, where males dominate females and white American culture dominates others. There are protections afforded in those traditional face-to-face channels of communication. In fact, some Net-watchers have observed that stripped of the social courtesies and contextual factors of traditional communication channels, this emerging communication environment is likely to be a less hospitable one than face-to-face (Adams, 1996). Consequently, more than replicating traditional imbalances in communication practice, this new environment also may not be comfortable for non-white, non-Americans, or females owing to its lack of contextual factors. In other words, the Internet may by nature be most conducive to the low-context culture of Western male society.

To this point, it is worth noticing that the African-American women participated more in class than on-line. In fact, one of the African-American women never joined the list (though it was required) and two only sent one message each. However, those women regularly come to class and two of them participated actively in class discussion. One of the observers remarked that it was if they felt "protected" in the classroom situation. This disparity in mediated communication environments has clear implications, then, as an increasing amount of business, educational and even personal communication is mediated.

While the lack of non-verbal cues is worthy of further examination, an even larger question emerges: is this technology actually fashioned after the values and perspectives of those who have created it (Rakow 1988; Spender 1995)? In the case of networked communication systems, the technology appears to be based on the dominant masculine value systems of Western society. In addition, there is different access to the creation of the technology as well; and the result has been the creation of a place were social practices extend the asymmetrical construction of power (Rakow 1988). Rakow states that key task, then, is to gain an understanding of how power is exercised through the technology.

With the growing use of computers, video games, and the Net, we may have a generation of children emerging within where everyone is equally comfortable in that environment. Also, the technology is changing rapidly, with dramatic increases in channel capacity, allowing for more and different cues to be included in the communication process. However, as the Net becomes "tiered," only some individuals will have access to this greater speed and broadband capacity, and the implications for those already underrepresented in those networked environments are enormous. Will they have the "left-over" old Net, or will they have the Super New Net that will allow them to have instant images, sound . . . maybe even touch and smell?

Perhaps the greatest implication is that in the near future, it is that the Net is still dominated in both academe and the business world by the discourse patterns of the dominant white male society, so that those women and men who want to participate equally, in the fullest sense (be heard, responded to, and part of the decision-making process of any negotiation taking place) will have to use the dominant patterns in order to "market" themselves. There are implications for policy from the most basic levels involving civility in the classrooms to the development and deployment of the GII.

While there is widespread recognition of the impact of new information technologies on individuals, institutions and society, there is little consensus on what that impact is and how and whether there should be remediation. However, that there is an information or technology gap is indisputable—and these obstacles to equity in information technology exist, most specifically for women, minorities and the poor.

Informed policy making has a role to play in addressing these inequities. Bowie (1990) states that it is the role of the government to protect the rights and interests of its citizens, especially the disadvantaged:

> What a government does for the human beings at the bottom of its social order—its poor, its minorities, its children, its women, its elderly, those who are underrepresented or unrepresented, as well as people who are handicapped, those who are undereducated, and those who are generally in need but cannot help themselves—defines the degree and quality of justice that can be expected in practice.

It appears, however, that the longer we wait to address this growing gap, the more likely we are to see these so-called "democratic" technologies contribute to an increase in inequity in participation worldwide, rather than to the emergence of an inclusive global information economy.

Notes

This manuscript appeared originally in the *Electronic Journal of Communication / La revue electronique de communication*, 8 (3 & 4), 1998 (see <http://www.cios.org/www/ejcrec2.htm>) and is reprinted by kind permission of the editors.

1. There were actually over 350 messages in total. However, the messages from the two instructors were eliminated from these analyses, since they primarily used the system for "broadcasting" (i.e., one-way transmission) of information relating to the in-class discussions.

2. The **Subject**: information in e-mail messages ends up not being very useful. It is often not indicative of a message content, especially when people simply use the reply function to send messages without revising the **Subject**: line.

3. For some of the Asians, there was a problem in using the English language.

4. Visual inspection of the messages revealed that most of the messages sent were only a few lines long and that there were far fewer longer messages. Given the paucity of data in this latter category, a simple classification scheme was most appropriate. Since messages less than half of a page consisted of little more than the header and a few sentences, messages were classified as long if they were more than half of a page in length.

5. For both figures, the amount shown in Week 11 represents the number of individuals who never adopted the listserv.

References

Adams, C. 1996. "'This is Not Our Fathers' Pornography': Sex, Lies, and Computers." In *Philosophical Perspectives on Computer-Mediated Communication*, ed. C. Ess, 147–70. Albany: State University of New York Press.

Balka, E. 1993. "Women's Access to On-Line Discussions About Feminism." *Electronic Journal of Communication* 3 (1). Available e-mail: comserve@vm.its.rpi.edu.

Bowie, N. 1990. "Equity and Access to Information Technology." *Annual Review of Institute for Information Studies*, 131–67.

Ebben, M., and Kramerae, C. 1993. "Women and Information Technologies: Creating a Cyberspace of our Own." In *Women, Information Technology, Scholarship*, eds. H. J. Taylor, C. Kramarae, and M. Ebben, 15–27. University of Illinois at Urbana Champaign: Center for Advanced Study.

Gudykunst, W, and Y. Kim. 1996. *Communicating with Strangers: An Approach to Intercultural Communication.* 3rd ed. New York: McGraw-Hill.

Hall, E. 1976. *Beyond Culture*. New York: Anchor Books/Doubleday.

Herring, S. 1993. "Gender and Democracy in Computer-Mediated Communication." *Electronic Journal of Communication,* 3 (2). Available e-mail: comserve@vm.its.rpi.edu. Message: Get herring V3N293

Herring, S., ed. 1996a. *Computer-Mediated Communication: Linguistic, Social and Cross-Cultural Perspectives.* Amsterdam/Philadelphia: John Benjamins Publishing Co.

Herring, S. 1996b. "Posting in a Different Voice: Gender and Ethics in CMC." In *Philosophical Perspectives on Computer-Mediated Communication*, ed. C. Ess, 115–46. Albany: State University of New York Press.

Herring, S., D. Johnson, and T. DiBenedctto. 1992. "Participation in Electronic Discourse in a 'Feminist' Field." Paper presented at the Berkeley Women and Language conference, University of California, Berkeley.

Hofstede, G. 1983. "The Cultural Relativity of Organizational Practices and Theories." *Journal of International Business Studies,* 15 (2), 75–114.

James, D., and S. Clark. 1993. "Women, Men, and Interruptions: A Critical Review." In *Gender and Conversational Interaction*, ed. D. Tannen, 231–80. New York: Oxford University Press.

James, D., and J. Drakich. 1993. "Understanding Gender Differences in Amount of Talk: A Critical Review of Research." In *Gender and Conversational Interaction*, ed. D. Tannen, 281–312. New York: Oxford University Press.

Kramerae, C., and H. J. Taylor. 1993. "Women and Men on Electronic Networks: A Conversation or a Monologue?" In *Women, Information Technology, Scholarship*, eds. H. J. Taylor, C. Kramarae, and M. Ebben, 52–61. University of Illinois at Urbana Champaign: Center for Advanced Study.

Lakoff, R. 1973. "Language and Women's Place." *Language and Society* 2: 45–79.

———. 1995. "Cries and Whispers: The Shattering of Silence." In *Gender Articulated: Language and the Socially Constructed Self*, eds. K. Hall and M. Bucholtz, 25–50. New York: Routledge.

Ma, R. 1996. "Computer-Mediated Conversations as a New Dimension of Intercultural Communication Between East Asian and North American College Students." In *Computer-Mediated Communication: Linguistic, Social and Cross-Cultural Perspectives*, ed. S. Herring, 173–85. Amsterdam/Philadelphia: John Benjamins Publishing Co.

Nadler, M., and L. Nadler. 1987. "Communication, Gender, and Intraorganizational Negotiation Ability." In *Communication, Gender and Sex Roles in Diverse Interactions*, eds. L. Stewart and S. Ting-Toomey, 119–34. Norwood, NJ: Ablex Publishing Corporation.

Nielsen Media Research. 1996. *CommerceNet/Nielson Internet Demographics Recontact Study*. Available: <http://www.commerce.net/work/pilot/nielsen_96>

Rakow, L. 1986. "Rethinking Gender Research in Communication." *Journal of Communication* 36 (4): 11–26.

Rakow, L. 1988. "Gendered Technology, Gendered Practice." *Critical Studies in Mass Communication* 5: 57–70.

Spender, D. 1985. *Man-Made Language*. London: Routledge.

———. 1989. *The Writing or the Sex (Or Why You Don't Have to Read Women's Writing to Know It's No Good)*. The Athene Series. New York: Pergamon.

———. 1995. *Nattering on the Net*. North Melborne, Australia: Spinifex Press.

Stewart, C. M., and S. F. Shields. 1996. "Women and Men Communicating in Cyberspace: Do Listservs Offer New Possibilities for Communication Equity?" Paper presented at International Communication Association conference, Chicago, IL.

Stewart, C. M., S. F. Shields, D. Monolescu, and J. T. Taylor. 1997. "The Dynamics of Online Chats: An Examination of Gender Issues in IRC." Paper presented at International Communication conference, Montreal, Canada.

Stewart, L., and S. Ting-Toomey. 1987. *Communication, Gender, and Sex Roles in Diverse Interaction Contexts*. Norwood, NJ: Ablex Publishing Corporation.

Tannen, D. 1990. *You Just Don't Understand: Women and Men in Conversation*. New York: William Morrow and Company.

———. 1993. "The Relativity of Linguistic Strategies: Rethinking Power and Solidarity in Gender and Dominance." In *Gender and Conversational Interaction*, ed. D. Tannen, 165–88. New York: Oxford University Press.

———. *Gender and Discourse*. New York: Oxford University Press.

Ting-Toomey, S. et al. 1991. "Culture, Face Maintenance, and Styles of Handling Interpersonal Conflict: A Study in Five Cultures." *Journal of Conflict Management* 2: 275–96.

Ting-Toomey, S. 1988. "Intercultural Conflict Styles: A Face-Negotiation Theory." In *Theories in Intercultural Communication*, eds. Y. Kim and W. Gudykunst, 213–35. Beverly Hills, CA: Sage.

———. 1985. "Toward a Theory of Conflict and Culture." In *Communication, Culture and Organizational Processes*, eds. W. Gudykunst, L. Stewart, and S. Ting-Toomey, 71–86. Beverly Hills, CA: Sage.

New Technologies, Old Culture: A Look at Women, Gender, and the Internet in Kuwait

Deborah Wheeler

Introduction

In 1994, a "comprehensive and reliable look at the user base of the Internet" found that nearly 90% of Internet users were male.[1] By 1996, studies suggested that one in four users of the Internet were female, an increase from 10% to 25% (Cherney and Weise, 4).[2] Jupiter Communications projects that by the year 2000, women will constitute approximately 44% of the on-line audience around the world.[3] What social impacts will result from women's increased access to information and communication capabilities? According to Eduardo Talero and Phillip Gaudette, consultants for the World Bank, access to new communications technologies can "raise cultural barriers, overwhelm economic inequalities, even compensate for intellectual disparities. High technology can put unequal human beings on an equal footing and that makes it the most potent democratizing tool ever devised" (Talero and Gaudette, 2). Will this be true for women? We are told that "wide spread networking coupled with the ease of publishing multimedia materials within the Web will support radical changes"[4]—but will the introduction of new media mean the creation of revolutionary new circumstances for women? Can women use new information and communications technologies to enhance their power and position in daily life? Lourdes Arizpe, Assistant Director-General for Culture at UNESCO, provides an answer when she suggests that "women can be on the front side of this revolution" and can use these new technologies to present their autonomous voices in the service of their own culturally diverse and regionally specific forms of liberation.[5] Such promises, however, deserve further investigation especially in light of women's daily realities in conservative Islamic countries like Kuwait.

How might new information technologies shape women's lives in Kuwait? In 1996–97 with a Council on the Foreign Exchange of Scholars Senior Post-Doctoral Research award, I studied the development of a Kuwaiti Internet culture and attempted to assess how Kuwaiti women were participating in these new communication opportunities. Throughout my research, my study of women and the Internet was continuously redirected towards the particulars of Kuwaiti identity and the social practices which regulate how the Internet can develop and spread, as well as what women do with the tool once they have access. While in Kuwait I learned that the expansion of Internet technologies does not take place in a vacuum. Given contextual factors, new information technologies will not necessarily promote democracy, economic growth, and improve women's lives, as many Western thinkers have argued.[6]

Careful ethnographic research of the development and impact of the Internet is best conducted at its point of application. This is especially true when studying gender issues and women's voices because practices on the ground are regulated by processes one cannot see from cyberspace. Grand theories about global and local restructuring in the wake of the Information Age often fail to consider that each country has its own culture, its own style of government, its own norms and sanctions on behavior and gender attitudes, its own socioeconomic status structure, its own level of literacy and education, and its own historical experience. These factors help to shape what kinds of communicative acts are enabled by the Internet, and which are discouraged.

The dialectic between new technologies and old culture is clearly at work in the case of Kuwaiti women and their use of the Net. Lived experience, including activity in cyberspace, is conditioned by processes of social regulation which do not fail to extend their reach into the spaces of human interaction enabled by networked communications. To demonstrate this point, this article first provides an overview of the evolution of a Kuwaiti Internet culture, and women's contribution to it. Second, this article records several examples of women in Kuwait narrating their relationship to the Internet, and uses these narratives as a window upon those aspects of culture and power which regulate women's daily lives. Third, this article examines some reasons why access to the Internet does not necessarily determine the result of use. In this section I discuss why the Internet's presence in Kuwait will not automatically revolutionize women's relationships to formal institutionalized power. While theorists in North America and Europe are fond of arguing that "for those in possession of information technology, power, influence, privileged status and domination are further enhanced and assured" (Acosta and Hartl 1996, 4)—women's lives in the Gulf suggest that advancement, even for those with access to the Internet, will continue to be contextualized in everyday forms of struggle and victory, which aim to carve out spaces for freedom in the face of deeply entrenched hegemonies of patriarchy.

Techno-Culture, the Internet and Women's Lived Experiences in Kuwait

With the money to buy new technologies and a culture which encourages their purchase, Kuwait has quickly adapted to the new information capabilities provided by the Internet. In 1997, Gulfnet International, Kuwait's main Internet Service Provider, estimated that at least 40% of the Middle East's Internet users reside in Kuwait.[7] Another survey conducted in 1998 by the Dabbah Information Technology Group (based in the United Arab Emirates) states that Kuwait has the highest density of Internet users per capita of any Islamic society.[8] Three main factors help to explain the vibrancy of Kuwait's Internet culture. First is the fact that Kuwait has one of the highest per capita incomes in the world, estimated at $23,300 (1997).[9] Second is that in Kuwait, new technologies and their acquisition are considered signs of social status. Third, supporting a culture of techno-consumerism, is the fact that the Kuwaiti government makes it a point to "get the latest technologies into the hands of all citizens as quickly as possible" as a means of distributing signs of affluence among citizens.[10] The importance of technological acquisition is seen in the local press, as special "technology" sections of local newspapers review products like digital cameras, laptop personal computers, new software, flat screen televisions, digital phones, and other techno-toys as soon as they are released by global manufacturers. This showcasing process is not just "eye candy," as might be the case in other developing countries; rather, each time a new product is reviewed in the press a list of retailers in Kuwait where such things are available for purchase is also provided. A "be the first on your block to have the latest technologies" attitude feeds Kuwaiti techno-culture and, subsequently, has made Kuwaitis anxious to get "wired," even if it means paying $10 an hour to use a networked computer at an Internet cafe.

When factors of age, education, and socioeconomic status are controlled for, women are just as likely, if not more, than men to use the Internet as a form of communication and entertainment. For example, in 1996, a survey of Kuwait University students found that roughly 55% of women surveyed considered the Internet a hobby.[11] This number contrasts significantly with regional data that in 1998 suggested that only 4% of all Middle Eastern women use the Internet.[12] Based upon survey data from Kuwait, it appears that age, literacy, level of education, and other socioeconomic factors may be more important than gender in determining who has access to the Web.

Internet access is growing in Kuwait, as the following chart suggests:

Internet Hosts in Kuwait

7/93 Hosts = 237

7/94 Hosts = 297

7/96 Hosts = 1,963

7/97 Hosts = 3,555

7/98 Hosts = 5,597

1/99 Hosts = 6,063[13]

A hosts count is less than a user count since often a host supports multiple users, like those host computers located at universities, private schools, businesses, embassies, even those in peoples' homes. Keeping the host number relatively low, compared with the United States, for example (over 1,000,000) is the fact that Gulfnet, Kuwait's only Internet Service Provider, charges approximately $150 a month for basic Internet service. Because of the high cost of service, most Kuwaitis, except for the very wealthy, do not have access from their homes. A large number of Internet users in Kuwait, however, access the Net from work, school, or one of the many Internet cafes in Kuwait. Exceptions include members of the Kuwaiti Engineering Society who can obtain an Internet account for around $30 a month through an agreement with Gulfnet. Moreover, every professor at Kuwait University is offered a "Slip" account free of charge, which enables dial-up access from home.

General consensus is that 95% of all Internet use in Kuwait is for "chatting." IRC (Internet Relay Chat) is the most common text found on computer monitors at cafes, university computer labs, and sometimes at work. These findings are confirmed by an informal survey which I conducted with the participation of university computer lab administrators at Kuwait University, Internet Cafe owners and employees, and IT industry professionals, parents, and students. These results were confirmed by a colleague who conducted her own independent survey at six Internet cafes in Kuwait, and found that only two respondents were using the Internet for "serious" business. An employee of Cafe Ole Internet cafe, located at the Layla Gallery, a high-end shopping mall, stated that "The Internet in Kuwait is about pleasure and down time in a climate where it's best to have the comfort of air-conditioning."[14]

Kuwait has the highest recorded temperatures for any inhabited area in the world. The country during the dry, hot summer months is also subject to frequent sand storms, which make Internet cafes a welcomed respite, especially for younger generations, both male and female. There are approximately ten Internet cafes in Kuwait City, including the Lady Di Cafe, established just months after the death of the Princess. It is customary for Internet cafes to be divided along gender lines—one side for females and one for males. These divisions are symbolic of the lines which separate men and women in Kuwaiti public life: at weddings, in lines at McDonalds, at University cafeterias, at government ministry waiting rooms, at library study lounges. These gender boundaries are policed by the eyes of a curious public and a strong sense of "you never know who might be watching" (a phrase I heard often throughout my fieldwork, mostly from women). Gender separation in public life is maintained by public fears of the cost for transgressing such boundaries; a cost usually assessed to a woman's, and thus a family's, reputation. The social sanctions against mixed gender interactions outside of direct relatives are so active that once while I was in an Internet cafe, the owner got a page on his pager. He called the number listed on his pager on his cell phone. He discovered that the page came from a woman inside the cafe. She had called to ask him to turn down the air conditioning as she was cold. She was sitting less than 20 feet from the owner, yet she did not feel comfortable communicating with him face–to–face, in a public place. When I asked the owner about this curious situation, he responded emphatically, "You know, gender issues."

In addition to being rich, mostly Muslim, conservative, advanced technologically, Net-active, and hotter than Hades for much of the year, Kuwait is a country where women face daily challenges because of their gender. Both in the press and in private conversations, the boundaries of patriarchy are tested by female social critique. Part of this discursive resistance is stimulated by the fact that women cannot vote and that women are denied many of the benefits provided to Kuwaiti men, like government-supported housing. A Kuwaiti woman is only guaranteed government housing through marriage to a Kuwaiti male. If she marries a non-Kuwaiti male, or remains single, she is unable to get government housing benefits. If she marries a non-Kuwaiti and has children, they are denied Kuwaiti nationality (and a whole host of government benefits) because nationality is determined by the husband in a marital relationship. For example, if a Kuwaiti male marries a non-Kuwaiti, he inherits all the same government benefits, and so do his male children. This arrangement puts pressure on women to get married, and to marry Kuwaiti men.

One professional, single (by divorce) mother whom I interviewed stated that even renting an apartment is a problem for unmarried women. In Kuwait, landlords do not want to establish a rental contract in a woman's name. This woman, who has a Ph.D. and is a mother, therefore had to ask her younger brother to sign for her. He also had to sign the contracts to sponsor a maid to care for her son, to get a telephone, and to buy a car. Single women who lack male relatives are severely encumbered. Some Kuwaiti women prefer to hold even abusive marital relationships together in order to avoid the social stigma and built-in difficulties of being single and female in Kuwait. One Kuwaiti woman who has studied gender politics in Kuwait summarizes the situation in this way:

> Women are still being persecuted for committing so-called "moral crimes." They have no legal protection against any form of abuse within marriage and no citizenship rights similar to those of Kuwaiti men, and face constant discrimination at work . . . Given the uncompromising stance of male society, it is clear that the challenge facing Kuwaiti women is daunting and changes will be slow to achieve. (al-Mughni 1993, 148)

Women in Kuwait, however, are much better off than women in many other Middle Eastern societies. In Kuwait, women can drive and are subject to compulsory public education from grades 1–12. Women attend universities and are commonly awarded government scholarships to study abroad. Women make up the majority of students in the Colleges of Medicine and Science, as well as the majority in the College of Education at Kuwait University. Thirty percent of the Kuwaiti work force is female, and of this female work force, two-thirds are married. In the government sector, which employs 92% of the Kuwaiti work force, half of the employees are women. The government guarantees employment to all citizens, both male and female, if they want to work. Marriage and/or gender does not necessarily preclude women from working, although there is an active public discourse trying to drive women into marriage, stating that "marriage is one of the signs and proofs of Allah in the universe" (al-Qaradawi 1997, 67). This same Islamic discourse encourages married women into the home stating that "the real place of the woman is in her home . . . raising children."[15]

Despite conservative public discourse, women in Kuwait (including married ones) are represented in all the professions, including medicine, law, academia, and business. Women are well represented in print and electronic journalism. The Journalists' Association recently elected a woman, Fatima Hussain, to its board of directors. She is also editor-in-chief of a prominent woman's magazine, *al-Samra*. Women are also a major part of the support staff that keeps complex government bureaucracies in Kuwait running, and the sense is that the society could not function without 50% of the small Kuwaiti society working (e.g., if all or most women stayed home). In 1996, Kuwait modified its labor laws to be more sensitive to women's needs and to meet international standards. This governmental action can be interpreted as a further entrenchment of women's presence in Kuwaiti public life. But while women are employed throughout Kuwaiti society, most of the leadership and upper-level management roles are reserved for men in both the private and public sector. In the words of one observer, "Whereas the West has a glass ceiling, in Kuwait it's concrete."[16]

While Islamics try to drive women back into the home, (or into fields "compatible with their nature" like education and nursing), liberal women propose counter arguments which stress that men need to share responsibilities at home because women are sharing responsibilities to provide.[17] At present, it is quite common in Kuwait to see fathers with their children at the store, at the movie theater, and having lunch out, although it is unusual to hear of men cooking or cleaning at home. In an interview with a middle-aged Kuwaiti woman it was noted that "younger generations of Kuwaiti men were more open to sharing responsibilities at home with their sisters, mothers and wives. Men forty and older would never be caught doing 'women's work.'"[18] When asked about the causes of change in the younger generations attitudes towards women and work, this woman observed that younger generations grew up with satellite TV, were more likely to travel and to study abroad, and thus were accustomed to different gender roles than their fathers and grandfathers. Might the Internet, like satellite TV, help to consolidate these small shifts in definitions of "women's work?"

Kuwaiti Networks and Women's Voices

One of the best ways to understand how the Internet is affecting women's lives in Kuwait is through oral testimony of the participants. The voices of women who are active Internet users reveal important characteristics regarding the cultural frameworks which regulate both women's lives and "networks" in Kuwait. Through these examples we obtain glimpses of the promise and problems of new communications technologies for women in the Arabian Gulf. These narratives were selected because of the women's differences in age, status, nationality, profession, and perspective, so as to provide a representative cross-section of the larger community of Kuwaiti women who use the Internet in their daily lives.

Nassima[19]

Nassima runs the Learning Resource Center at a private school in Kuwait. She is middle-aged, a Kuwaiti citizen, and a self-taught computer technology expert. Every day, she introduces new Internet users to the tool's power. Yet her words remind us of the ways in which the Internet can reinforce boundaries between genders, if not deployed in ways compatible with women's lives in Kuwait. I visited her at the school in the last week before classes let out for summer recess in June of 1997. We spent an hour together talking about the Internet, education, and gender issues, as well as viewing some of the educational materials provided by the school to guide Internet use. We laughed together when we viewed the "bookmarks" student's maintained on the center's Netscape-based Web browser. The bookmarks suggested that the Internet connection at the school was a tool for male pleasure, rather than female gender education/resistance. She observes:

> Girls don't use the Internet unless required to for a class. Boys come after school and use it for pleasure. They go to

> sites with cars, sports, pop music. The only time girls got actively involved was when they were using the Internet for horoscope information. Girls have a different attitude towards technology than boys. Boys are comfortable with it and like to play with it. Girls are not comfortable with it and would much rather giggle together and talk. Boys teach other boys how to use the Internet and how fun/useful it can be. Boys don't teach girls for obvious reasons, and few girls, if any, are highly skilled in the technology and able to teach others. Thus girls don't learn to be comfortable with the technology in a non-threatening way. Girls are expected to go home after school. Boys are able to come after school to the center to play with the Internet. Once a boy tried to access a site on explosives. A message appeared on the screen, "You are forbidden to go here." Forbidden by whom, to this day we still do not know. The Ministry of Information censors our Internet guides. Here, look at this Web magazine. The cover advertised a story that discussed love on the Internet, and looked at how boys and girls are developing relationships through the IRC. The Ministry censored this.

Nassima's narrative on the surface seems to reinforce stereotypes about girls and their relationships with technology and science. She notes that girls are unlikely to use the Internet in their free time, and are unlikely to teach other girls how to use it. These conclusions contrast highly with the fact that when young Kuwaiti women attend the university, large numbers of them choose to major in fields like computer engineering and pre-med, and perform well. At Kuwait University, for example, there are more women than men in science and medicine programs, and female students in the sciences continuously outperform male students in terms of grades and test scores. In 1997, for example, Kuwait University raised the test scores required for female entrance to the Engineering department, while they lowered standards for male students, because women were scoring better on entrance exams than men and the school wanted to balance enrollments. The reason given was that administrators wanted to avoid placement problems for graduates, stating that the market couldn't support large influxes of female labor.

Nassima's own story of being a self-taught Internet expert, very comfortable with the technology and able to teach girls how to use it, challenges the narrative she herself constructs about the Internet and gender. When Nassima's narrative is viewed in light of the testimony of other women interviewed for this article, it becomes clear that Internet use at this particular school is influenced by the administration and architecture of the lab, and thus not necessarily representative of young Kuwaiti women as a whole. For example, at the Learning Resource Center, the only time for free use of the Internet is after school, when girls are most likely not able to "stay after." Families in Kuwait tend to keep track of their girls in a way that they do not track boys. Girls have a clearly defined place within the home and the extended family network to which boys are not as strictly subject. Perhaps if there were Internet free play hours available for girls during the regular school times, they would be more apt to play with the technology. I have found that when homes have an Internet connection, girls are just as apt as boys to use the tool. Both boys and girls tend to gather with friends at home or at Internet cafes to surf the Net. The physical layout of the Learning Resource Center also limits female use. For example, there is only one computer at available with an Internet connection. If boys are using this machine, girls are unlikely to play along with them. As discussed previously, there are active cultural hegemonies in Kuwait which keep genders from mixing. Boys and girls do not feel comfortable sitting together, thus with one computer to use in the Learning Resource Center, girls are not as likely in this environment to integrate Internet use into their social practices. Perhaps if the learning resource center had separate hours "for girls only" during regular school hours, they would be more likely to learn and to play with this tool.

Su'ad

Su'ad is an electrical engineering student at Kuwait University. She is twenty years old. She began using the Internet in college, and admits that at one point she became addicted to it and had to quit cold turkey for several months until she got her use under control. Now she limits herself to access once a day. Her narrative is important because it provides a contrasting image to that constructed by Nassima. She explains:

> I use the Internet every day. I come to the lab and use IRC. My little sister uses it too. She's been using the Internet since she was six. She's eight now. People hack around all

> the time. Here in Kuwait, many people use other people's accounts to surf the Net. They break into places where they're not authorized to go. I hack. The Internet cafes are full of users. Have you ever been? I'll take you there and teach you to use IRC. In Kuwait, if you give people freedom they will misuse it. This is why the Internet is dangerous. Kuwait channel 1 on IRC is all about sex. I prefer Kuwait channel 2 instead. I meet interesting people there. One man who is Kuwaiti but is studying now in London has been pursuing me on IRC. He sent me his picture as an uploaded file through IRC. I got it and I started laughing. He looks just like my father. I could never marry him. I want someone very handsome. I told him this and he said to give him a second chance because the picture was not really a good one . . . I have a friend who is getting married to someone she met on the Internet. They only "chatted" for four months and now they're going to spend the rest of their lives together. I think she is stupid. It's possible to lie on the Internet. How does she know that he is really as good as he says he is on-line. One has to be careful. . . . One time I was "chatting" with another engineer from Saudi Arabia. He kept asking are you a man or a woman. Finally I answered, "I'm a woman, is this important." He said, "Yes, I refuse to talk to you."

Su'ad's narrative provides an image of a woman at ease with technical environments. Her words are representative of the many young women at Kuwait University who are specializing in the sciences and are serious about their advancement within Kuwaiti society. Many young Kuwaiti women major in the sciences because their chances of employment in the medical and scientific fields are high.[20] Like many of these young women, Su'ad wears a veil, but this seems like an almost irrelevant detail. Many women veil because of the anonymity it provides them in public places. Being veiled can enable women to meet and speak with members of the opposite sex: this would otherwise be difficult if everyone knew who they were. Su'ad explains, when asked about veiling, that she is respective of her Islamic values, yet she observes that she was raised in a liberal environment by parents who were educated in the US. Thus she feels comfortable in crossing strict gender lines on IRC and in real life. For example, once when Su'ad took me to an Internet cafe, she said that someone she met on IRC would be meeting us there. "A guy?" I asked. "Yes," she said. "A guy. Don't worry, I know he's okay. I asked my friends about him and they said he was worth meeting." I was surprised by her boldness, because of the hegemonies regarding gender that had been at work upon the lives of many women I had met in Kuwait. When we met this at the cafe, we found out that he was in his twenties and was an electrical engineer who worked as a troubleshooter for Kuwait Airlines. He joined us at a table, and so did the cafe owner. The four of us sat and talked about how the Internet was changing Kuwaiti society. Our conversation was perhaps symbolic of the broader changes taking place in Kuwait. The computer brought us together, the computer cafe provided the context, and new freedoms of trans-gender interacting in cyberspace made us all comfortable sitting and conversing face–to–face.

Su'ad's willingness to teach me how to use IRC challenges Nassima's observation that girls don't teach others how to use the Internet. Her willingness to help guide me through IRC's special linguistic codes reveal concretely how the education process works. In return, I have also been asked by women in the labs at the university to help them when they are just learning IRC. I'm not an expert, and often there is a male student who is sitting right next to them. Local hegemonies do not enable them to ask males for help. The more women who are comfortable with new communications technologies, the more examples there are to follow, and the more potential teachers there are who can help women get on-line.

While for some users, getting on-line in Kuwait means entering a whole new world where men and women learn to interact with the other in non-threatening and previously inadvisable ways—these interactions continue to be conditioned by the local codes of a conservative Islamic environment, symbolized by the Saudi man's refusal to even "chat" with Su'ad on IRC. The advent of the telephone, the cell phone, the shopping mall, the automobile, all of these innovations have not rendered benign the effects of conservative Islamic culture on Kuwaiti lives. Observers should not expect the introduction of a tool like the Internet to do so either. Genders mingle on IRC chat while men and women sit segregated in Internet cafes. To mingle in cyberspace is safe, to do so at the mall or on the street is not. Thus in general, new technologies are adaptable to local environments, and usage conforms so as not to cause open and offensive violations of local cultural codes. In Kuwait (and other places such as Singapore), an equilibrium exists in terms of Internet use between "permitting room for creative expression and maintaining society's moral standards" (Low 1996, 12).

Badriya

Badriya is a computer science major at Kuwait University. She is nineteen and originally from Iran. I interviewed her in the computer lab at Kuwait University. Badriya is conservative and veils but she, like Su'ad, has liberal, outspoken ideas. Her narrative emphasizes the ways in which the Internet is changing women's status, at least in cyberspace. Her words also remind us of the contextual hegemonies that prevent use of the Internet for open and active gender resistance.

> I use the Internet daily. I use it mostly for entertainment purposes, when I get bored, which is often. There's not much for young people to do to relax in Kuwait; the Internet fills this social gap. One of the things I like most about the Internet is that it allows girls to speak with authority, whereas in real life, men constantly tell women that men are superior and that women shouldn't speak. Still, I don't think that the Internet will support active struggles for women's rights. Politics are dangerous here. People are afraid to speak. If you're an important person, or a person with connections within Kuwaiti society, you can say whatever you want. If you're a small person without public importance you cannot; you lack protection. Most women lack protection. Even some important people are afraid to speak in Kuwait. People can take what you say the wrong way and use it against you so most people just maintain a low profile to protect their reputations.

Badriya's narrative emphasizes once again the ways in which the Internet has the power to change woman's voice, by de-emphasizing the gender of the speaker. In spite of these changes in women's voice online, such narratives remain contextualized in, for example, a legal system where, under certain circumstances, a woman's testimony only counts for half a man's testimony. Kuwait is a society where, a century ago, houses were designed so that women's voices would not be heard by visitors. One Kuwaiti historian elaborates by observing that "it was considered *aib* [shameful] to let women's voices be heard." Historically, the policy was that "women should not be seen or even heard by men who were not relatives" (al-Qinanie 1968, 66). These attitudes towards women's voices still have an impact upon female voice today. There are unwritten rules governing when a woman should speak publicly. Given the stress on marriage as a social given, young women are careful about how they behave, conscious of their heritage and culture, and how outspokenness may be interpreted by potential suitors. One result is that in college courses at Kuwait University where I occasionally guest lectured, women are highly unlikely to speak in class, even though they usually end up with better grades, which means that their silence is not about lack of understanding. I was told many times during my research in Kuwait that the main difference between Western women and Kuwaiti women is that Kuwaiti women, even so-called liberated ones, would be unlikely to act or speak in a way disapproved of by their husbands and male relatives, whereas women in the West were free to act and to speak as they pleased. Thus, although gender neutrality provided by some transactions in cyberspace enables some men and women to learn about each others' lives, these cyber-liberations remain constrained by history and local culture on the ground.

Interlude: Power and Voice in Kuwait

Badriya's narrative also reminds us of the power constraints upon people's voices in general. If one is not a person with *wasta* (an Arabic term for "connections"), then one is not protected from the potential harms of speaking out. Women tend automatically to have less *wasta* than men. Even those who are from prominent families are very careful about what they say. Throughout my research in Kuwait, women would utter, "Don't quote me on this," "Off the record of course. . . ." Most women who were cautious about being quoted had some story to tell which revealed why they were concerned. One woman, who has a Ph.D. and comes from one of the leading families in Kuwait, explained that a man she knew spoke out against an Islamic member of parliament, and as a punishment lost his summer teaching opportunity. Thus, she notes, "We need to be careful."

Another woman, also a Ph.D., noted that she once read an interview conducted by *The Observer* with two Kuwaiti women studying in London. The interview touched on sensitive issues such as virginity, women's honor, marriage, and women's rights. My friend was impressed, and scared, by these two women's boldness. She had read the published interview in *The Observer* while she was studying abroad. While talking with a friend some years later on the issue of women's voices, my friend recounted her amazement at these two women's boldness in *The Observer* interview. Her friend said, "Yeah, you remember what happened to the one who was not protected by *wasta*!" My friend said no, and explained that she was out of the country when this article was originally published. Her friend explained that as soon as word of the published interview reached Kuwait, it stirred quite a scandal in the local community, and both the girls' reputations were blotched. The mother of the less "protected" of the two students feared that this tarnish on her daughter's reputation would prohibit her from being "suitably married." The mother flew immediately to London, and discovered that, evidently, the student with "protection" (by her status) encouraged the other one to participate in the interview. As compensation for her irresponsibility and insensitivity, the mother demanded that her daughter be married to one of the family's eligible sons. Several months later, the less protected young woman was married, whether she wanted to be or not.

Another of my friends explains that during the Gulf War, she and a friend gave an interview to a Western newspaper describing women's activities against Saddam's occupation forces in Kuwait. They gave the interview based upon conditions of anonymity. But the reporter revealed so much personal information about the two women that "everyone" in Kuwait knew it was them who had spoken publicly about issues many Kuwaitis wanted to keep private. Ever since this experience, this friend of mine has kept a low profile.

These are examples of the kinds of stories (perhaps urban legends?) that are told to reinforce boundaries between public and private discourse in Kuwait. The attempted murder of a Kuwaiti colleague of mine and her husband as they drove home from their chalet one evening in the spring of 1997 reveals the costs of being outspoken even for the very well-connected. Both she and her husband have rich public records of organizing for women's rights (the wife, an outspoken woman's activist) and fighting corruption in the misuse of public funds (the husband, a veteran MP). The sense was that this murder attempt was directed at the husband for his campaigns in parliament to oust those government ministers who were robbing the public of their livelihood, and who were looting the nest egg of future generations. Evidently he got too close to embarrassing some very important people in public and they tried to permanently remove him from office. This event took place in the middle of my fieldwork and was very disturbing as it revealed the degree to which some individuals would go to keep things quiet. Kuwait on the surface is a democracy, with a very free press. Members of parliament are very outspoken in the challenges they make, even during open sessions, against the government. Yet simmering beneath the surface is public fear of going too far in one's discursive activities. These narratives of "punishment" for speaking too frankly, and too publicly, represent several examples, some minor, some extreme, which police patterns of public behavior in random ways. Kuwait is not a police state: it is rather a very small community where every one is related in some way, and where strength of the community is valued over the right of the individual to speak. If corruption exists, which it does in every family, then sometimes it's best not to discuss it openly, so that from the outside, appearances suggest that all is well. The status quo is valued over change. Anyone daring to speak out for change has to be prepared for a whole range of possible consequences.

Because voices are constrained by social sanctions against speaking out, the majority of individuals in Kuwait will not feel comfortable using the Internet to publish information which could be used against them in the "social courts" policed by their neighbors, relatives, employers, and friends. Cyberspace is an extension of the realms of social practice and power relations in which users are embedded. At times voice is liberated from gender restrictions, like from within the cyber-relations enabled by IRC. But voice is historically subjected to constraints based upon publicly enforced notions of right and wrong in public discourse. The advent of new fora for communication do not automatically liberate communicators from the cultural vestiges which make every region particular and which hold society together. In Kuwait, this means that women are not likely to organize and to speak out against their husbands, their brothers, their sons and fathers, their bosses; this would be to publicly embarrass their patriarchs. It is more likely that voices will be lifted in the privacy of an office with the door shut, or in the living room during hours that many husbands are at work. In my experiences, these voices will cut to the quick of the matter, will express a well-reasoned and culturally-seasoned opinion of women's lives in the conservative Gulf. These voices, if uttered by liberal women, might stress that men in the Gulf are simply afraid of women and their power, and are unlikely to yield to women's desires if overt and confrontational demands are made. These women stress that women's struggles for liberation in the Arab world require subtlety, and compromise, rather than all-out revolution. Seduction and charm are the best tools for carving out spaces for women's freedom. If the voices come from conservative Muslim women, they are likely to point to the kinds of oppression women in the West face, which women in the Islamic world are free from: the compulsion to work, an inability to depend upon men to provide, and the public exploitation of women's bodies. Regardless of what form expression takes, the main point is that women's narratives are more comfortably distributed in face-to-face conversations, in privacy, in ways that keep information within the circles for whom it is intended, narrowly defined. Public trust in the privacy of the Internet is not present and questions about government monitoring or a lack of anonymity are lingering in any user's mind.

Layla

Layla is a prominent Kuwaiti woman in her late fifties. She is originally from England and is Kuwaiti by marriage. She has been subjected to the pressures of the family matriarchs' reserve for "foreign imports" into the bloodlines. These experiences have made her more vocal about women's issues in Kuwait, and her voice has not been one to advocate their need for liberation. On the contrary, Layla's narrative describes the power and control women already have over Arab society. In her words, "Women in Kuwait aren't in need of any more power. It is in their nature to want to control everyone and everything around them. Arab women are strong, and many are mean. I'm afraid of them." I heard versions of these observations from many Kuwaiti men. One man told me that "You don't understand the distribution of power in Kuwaiti society if you think women needed any more power. They already control society." Another told me that Kuwaiti "women have traditionally managed social and financial day-to-day life in Kuwait, because of the heritage of pearl diving. Men went away for months at a time to dive. Women were left behind to take care of the community." Layla and these other male voices encouraged me to look beyond the rhetoric of patriarchy that on the surface concealed women's power. Underneath these discursive chains lies a world which women rule. Women control the lives of men, and they run the country from behind closed doors. Cartoons in the Arab world commonly poke fun at this relationship by representing a large, strong, powerful woman, hovering over a cowering male. The caption commonly has the woman making some demand of the man, and the man rarely is allowed to disagree.

My visit with Layla took place in her lovely home. I begin by telling her about my research and how I hoped she could help me to understand Kuwaiti women's lives, and the ways that new communications technologies like the Internet might change them. My first question asks her whether or not she sees the Internet as a positive force for women's empowerment. She responds:

> Well I don't think women in Kuwait will use the Internet for positive social change. They are lazy and would rather talk about superficial things like make-up and fashion. Women are also inhibited in what they say publicly to protect their reputations.

The critical voice behind this narrative in part attacks public standards of women's appearance in Kuwait which stress the value of a woman in terms of her beauty. Women in Kuwait face great pressures to conform to standards of appearance, many of which are defined and maintained by other women. After all, it is mostly women who see other women unveiled. It is mothers who look for potential mates for their sons at wedding receptions and parties where large groups of young women appear unveiled. Several times I was told by other women that I was "getting fat, and better watch out because my husband would be unhappy." I was told by some women that I should buy more expensive clothes, or wear more make-up because the impression someone makes through dress and appearance is very important in this part of the world where men run things. One man, before a meeting he set up for me with a powerful member of the ruling family, asked me if I had to wear my glasses, and told me to dress in an appealing way, not to wear my "ethnic jewelry." He asked me to remove my glasses to see how I looked without them. He said, "yes that's better, see how beautiful you are without them." He said that he was giving me advice as a friend, that Arab men liked to have admiration from beautiful women. I told him that I didn't want to date this person, I wanted a professional relationship. He said, "It doesn't hurt to use your sensuality to get what you want." I interpreted his words as sexual harassment. I interpret the pressures from other women towards conformity as an attempt to put chains on my individuality, my cultural difference that stresses natural appearance and casual dress (Southern California upbringing) and mental capacity over looks (academia). One woman told me that she would much rather have a male boss than a female one, because women were "back-stabbing, jealous, and unstable. Men were predictable and easily manipulated by women's ways." Such sentiments help to reinforce structures of patriarchy and women's subordinate place within it.

During our conversation, I tell Layla about my experiences with Nassima, the head of the Learning Resource Center, and discuss girls' relationships with the Internet in the private school I visited. I share with Layla my theory about Internet "free time" and its inaccessibility for girls at the school. I ask her about the family constraints placed upon children. I ask her about her interpretation of girls' freedom of movement and association and whether or not she thinks girls are more encumbered than boys. She observes:

> I suppose that's right. I guess girls are subject to different expectations than boys. The truth is that there is pressure on all family members to be at home together for the midday meal. Afterwards everyone takes a nap. If boys wanted to skip this meal, then their families would probably overlook their absence and would explain the situation in terms of boys needing solidarity with their peers. Girls, however, need to express their first loyalty to the family and are expected to be at home, protected and safe.

"And to help with the dishes," I muse, remembering my own childhood. She laughs:

> No, Kuwaiti women and girls do not help at home. They have maids to take care of domestic responsibilities, to watch the children. It's just the pressure to be home together to which girls would be held more strictly to than boys.

Layla is one of the only women I have found whose organization has a home page on the Web. I ask her about it and she tells me that she "helped design it." I want to ask her to what degree she helped, but I dare not, lest I loose my place holder in an almost empty category of Kuwaiti women developing content on the Web. Her response that "she helped design the home page" is probably more representative of a woman with the resources to hire another to do the labor-intensive work required to produce content on the Web, rather than illustrating a woman afraid of technology.

When asked directly about her understandings of the implications of the Internet for Kuwaiti society she states:

> I'm worried about what the Internet will do to Kuwait. First of all, this is a society which is not prone to read. The Internet, like satellite TV and video games before it, further encourages Kuwaiti youths to avoid reading books. I'm not sure that they will use the Internet for serious research, as

> they lack the skills to search for information that is not easily accessible through personal association. I'm most worried about how it is changing youths' attitudes towards sex. We're seeing it schools now. Students are more comfortable interacting across gender lines than ever before. Young people are experimenting with their sexuality in ways not common to this conservative society. Adults would rather close their eyes to these changes and pretend that traditions live on. These are circumstances in which youths will not get the information they need to be safe in their experimentation, that is, until there is a problem which is out in the open and publicly acknowledged, one which cannot be concealed. This same process happened with the drug epidemic.

Layla's narrative helps to reinforce the images, presented in a more sympathetic voice, by Su'ad and Badriya. Su'ad and Badriya are both part of this new youth sub-culture, which Layla characterizes as "interacting across gender lines . . . experimenting with their sexuality in ways not common to this conservative society." To a degree, the Internet is helping to support this culture of openness towards new gender values. But public sanctions on such openness still remain the norm; and it is in light of conservative morals that right and wrong are judged, keeping to a limit the degree of impact which Internet technology can have on young peoples' lives.

Lessons from the Kuwaiti Case: Culture, New Technology, and the Persistence of Local Values.

In my studies of women's networks in Kuwait I have tried first to understand the culture which surrounds and regulates women's lives. I have next tried to interpret how new communications technologies and their use in Kuwait make transparent local cultural practices. From within this context, I have then tried to grasp how the Internet is shaping women's lives. In Kuwait, local cultural constraints make female Internet use a limited force for social change. Most women with whom I spoke considered the Internet to be a tool for global feminist practices, divorced from, and unable to help with the local struggles of Kuwaiti women.[21] Cultural hegemonies in Kuwait, which define women's place, women's voice, and women's activism, limit open and organized gender struggle. The need to publicly link a text to a voice means that social pressures towards conformity for the good of the "family" (broadly defined) acted more powerfully to constrain women's voices than to liberate them. In the case of Kuwait, a woman's reputation, and the ways that local information can be distributed to harm it, creates an institutionalized pattern for women's activism and voices. Even in an age where the Internet offers women great freedoms of information and self-representation, women in Kuwait, including activists for social and gender equity, view local problems as better solved by a "kitchen cabinet" (or, more appropriately for Kuwait, a parlor committee).

Just because the Internet in Kuwait is not supporting open, organized, and sustained feminist resistance does not mean that women are not an important part of Internet culture in Kuwait, nor that their cyberactivities are not having any social impact. As Carolyn Marvin observes, "Electric and other media precipitate new kinds of social encounters long before their incarnation in fixed institutional form" (Marvin 1988, 5). This statement could easily be applied to the emergent Kuwaiti Internet culture and women's place within it. On the one hand, we are not yet seeing the emergence of new "fixed institutional" relationships for women within Kuwaiti society (like the vote) at the advent of the networked era, as so many Western social critics predicted would generally be the case. Kuwaiti women have yet to use their new communications possibilities "to put themselves on equal footing" with men (Talero and Gaudette 1996, 2). Yet the Internet is supporting a whole range of "new social encounters" in cyberspace. The Internet in Kuwait is enabling women to go to new places (like chat rooms where they can converse unescorted with members of the opposite sex) and to speak without having their gender influence the response of the conversant. It is enabling young couples to meet and choose mates independent of the family patriarchs. The Internet is opening new employment opportunities to women like Internet cafe ownership, or home page design. These incremental changes could result in significant social gains for women in Kuwait in the near future; but for now, cyberpractices continue to be shaped by the givens of Kuwaiti culture, until the result of this dialectic between old culture and new technologies takes root in local landscapes, producing an institutionalized synthesis.

Conclusions: Kuwait and Women in the Larger Muslim World

My research in Kuwait suggests that the age of the Internet could bring increased gender equity for women by providing them new professional opportunities sensitive to the cultural constraints of Islam, and as well, improve relationships between men and women, as greater intimacy outside and before marriage is possible through Internet chat lines. These "liberations," however, occur against a background of continued conservatism both politically and culturally, which limits overt "revolutionary" impacts of the Internet on women's lives. The stability provided by the regulatory systems of Kuwaiti society, although they limit overt feminist activism, at the same time provide some semblance of organization, and give Kuwaitis rules, both spoken and unspoken, of local protocol. The uncertainties caused by the Iraqi occupation, and the social turmoil which was and is its legacy, keeps many feminist activists more focused upon processes of social healing than upon their own "self-centered" advancement. The increase in drug abuse, divorce, violent crimes, unemployment, moral crimes, and the unhealed wounds of the Iraqi occupation, such as POW issues and continued complications of post-traumatic stress syndrome, help to divide women's solidarity along social activist lines. The luxuries of life that many Kuwaiti women enjoy further divide women among lines of haves and relative have-nots, and divides as well the haves along lines of beauty and appropriate social presentation. Differences in degrees of religious observance, as well as sectarian differences also divide the power of women's voice. Thus, one comes to understands women and the Net in Kuwait against a complicated local background of cultural, social and political givens, which shape and limit activism along many lines presenting a chorus of women's voices, each singing for themselves, and their community, many according to different tunes.

The case of Kuwait shows how activism is shaped by local institutional and cultural imperatives, factors which discourage the majority of women in Kuwait from openly testing the chains which male hegemonies provide them. This case illustrates that just because capabilities to "know" and to "speak" are provided by the onslaught of new communications tools does not mean that such tools will be used freely, without contextual constraints. Rather, a complicated context of cultural, political and social institutions weaves itself around women's use of information in conservative Islamic societies. It is important to understand the lives and voices of women throughout the world lest we make the mistake of thinking that "access" is the primary issue in building global solidarity with feminist consciousness and activism. Only by understanding the constraints upon women's activism across the world can we know where to find, how to interpret, and how to encourage women's voices in the new electronic frontier of which many of us find ourselves a part.

Notes

1. <http://www.cc.gatech.edu/gvu/user-surveys-09-1994/html-paper/survey>

2. For more on the relationships between women and computers see Spender (1995), Grundy (1996), Harcourt (1997), Chisholm (1996). Examples of relevant Web sites include: "Women Active on the Web" <http://www.Web-publishing.com>, which showcases select home pages developed by women; and "Virtual Sisterhood" (a site to help women get net active), <http://www.igc.apc.org/vsister>.

3. <http://www.nua.ie/surveys/?f=vs&art_id=904318494&rel=tru>

4. Talero and Gaudette (1996, 2). For more surveys of women's global Internet usage patterns see <http://www.nua.ie/surveys>.

5. Conversation with Lourdes Arizpe, World Bank meeting in Toronto, May 2, 1997.

6. For example, see Rheingold (1994), Heilemann (1997), Nye and Owens (1996). Characteristic of this optimism is the claim that "Everyone benefits, particularly the underdeveloped economies, which take advantage of the leapfrog effect, adopting the newest, cheapest, best technology rather than settling for obsolete junk" (Schwartz and Leyden 1997, 129). For arguments regarding the positive social impact of the Internet on women's lives, see Ebben and Kramarae (1993).

7. These statistics were part of a display constructed by the company at the Info World '97 trade show, January 1997, in Mishrif, Kuwait.

8. Survey results published in *The Star*, 23 July 1998. <http://star.arabia.com/980730/TE2.html>

9. *CIA World Factbook*, 1998. On-line version: <http://www.cia.gov/cia/publications/factbook/ku.html>

10. *Arab Times*, 15 March 1997, p. 1.

11. Survey administered by Dr. Saif Abdal-Dehrab Abbas, Professor of Political Science, Kuwait University. Results analyzed by the author. This survey was administered by students in a Methodology course conducted by Dr. Saif. The fact that it was conducted by Kuwaiti students on Kuwaiti students increases the reliability of the results. Westerners who have attempted to collect survey data in the Middle East have met many

frustrations, some of which are avoided when relying upon locals for implementation, as well as student populations as subjects, students being more apt to answer frankly. For more on this subject see, O'Barr et al. (1973), especially the sections by Mark Tessler.

12. <http://www.nua.ie/surveys/>

13. These figures were obtained from two sources: Network Wizard's Web site <http://www.nw.com> and personal e-mail correspondence with Grey Burkhart, CEO Allied Engineering, g.burkhart@computer.org, who provided the 1/99 data from his forthcoming publication, "15 January 1999 GITAG Survey of the .kw Domain."

14. For more information see the cafe's Web site at <http://www.ole.com.kw>; or e-mail the cafe directly at webmaster@ole.com.kw

15. "Kuwaiti Women Strongly Resent Idea of Staying at Home." *Arab Times*, 8 January, 1994.

16. A comment made during a conversation with a senior female member of the American diplomatic mission in Kuwait.

17. See for example, *al-Samra* (February 1997), and *Muntada al-Marah wa Sana' al-Qirar: Bahath wa-Awraq al-Amal* (Kuwait: Women's Cultural and Social Society, 1996), 61-73.

18. Interview with a working woman at the Ministry of Education, 8/6/97.

19. Note: the names of the women interviewed here have been changed to protect their privacy.

20. If one looks at the employment classified ads, those calling for high-tech jobs rarely specify job qualifications along gender lines. By contrast, jobs which require driving and selling, or are management related are likely to use gender-specific language in their advertisements for the position.

21. When pressed on this issue, several women responded that in Kuwait, the women who needed to be reached the most with regards to their rights were not computer literate: thus the Internet would do little to help them. Rather, according to a number of women's rights advocates I interviewed, the best channels of communications for campaigns to increase women's access to social justice in Kuwait would include: hotlines for domestic abuse; special counseling offices for dealing with marital problems; social agencies dedicated to dealing with post-traumatic stress problems created by the terrors of the Iraqi invasion; and pamphlets printed in simple Arabic explaining women's legal rights under the *Sharia* [the tradition of Muslim law] (such as the fact that women can write into the pre-nuptial agreement that a man cannot take another wife).

References

Acosta, Olivia, and Mari Hartl. 1996. "Women and the Information Revolution." *Women 2000*, 1 (October): 1–6. New York: United Nations Division for the Advancement of Women. <http://www.un.org/dpcsd/daw>

al-Mughni, Haya. 1993. *Women in Kuwait: The Politics of Gender*. London: Saqi Books.

al-Qaradawi, Yusif. 1997. *Makanah al-Marah fi Islam* [*The Status of Women in Islam*]. Cairo: The Islamic Home Publishing Co.

al-Qinanie, Yusuf. 1968. *Safahat min Tarikh al-Kuwayt* [*Pages from Kuwait's History*]. (Kuwait: Government Printing House. Quoted in al-Mughni, 42.)

Cherney, Lynn, and Elizabeth Reba Weise, eds. 1996. *Wired Women: Gender and New Realities*. Seattle: Seal Press.

Chisholm, Patricia. 1996. "Cyber-Sorority: Women Begin to Feminize the Net." *Maclean's*, 109 (Nov. 18): 53–54.

Ebben, Maureen, and Cheris Kramarae. 1993. "Women and Information Technologies: Creating a Cyberspace of Our Own." In *Women, Information Technology and Scholarship*, eds. H. Jeanie Taylor, Cheris Kramarae, and Maureen Ebben, 15–28. Champaign-Urbana: University of Illinois Press.

Grundy, Frances. 1996. *Women and Computers*. Exeter: Intellect, Ltd.

Harcourt, Wendy. 1997. *An International Annotated Guide to Women Working on the Net*. Geneva: UNESCO/Society for International Development.

Heilemann, John. 1997. "The Integrationists vs. the Separatists." *Wired* (July): 53-56, 182–87.

Low, Linda. 1996. "Social and Economic Issues in an Information Society: A Southeast Asian Perspective." *Asian Journal of Communications* 6(1): 1–17.

Marvin, Carolyn. 1988. *When Old Technologies Were New: Thinking About Electric Communication in the Late Nineteenth Century*. Oxford: Oxford University Press.

Nye, Joseph S., Jr., and William A. Owens. 1996. "America's Information Edge: The Nature of Power." *Foreign Affairs* (March/April): 1–32.

O'Barr, William, David H. Spain, and Mark A. Tessler. 1973. *Survey Research in Africa: Its Application and Limits*. Evanston: Northwestern University Press.

Rheingold, Howard. 1994. *Virtual Communities: Homesteading on the Electronic Frontier*. New York: Harper Perennial.

Schwartz, Peter, and Peter Leyden. 1997. "The Long Boom: A History of the Future 1980–2020." *Wired* (July): 115–29, 168–73.

Spender, Dale. 1995. *Nattering on the Net: Women, Power and Cyberspace*. Melbourne: Spinifex Press.

Talero, Eduard, and Phillip Gaudette. 1996. *Harnessing Information for Development: A Proposal for a World Bank Group Strategy*. On-line version: <http://www.worldbank.org/html/fpd/harnessing>.

Preserving Communication Context: Virtual Workspace and Interpersonal Space in Japanese CSCW

Lorna Heaton

This paper describes the design of systems for computer supported cooperative work (CSCW) in Japan with particular attention to the influence of culture. In doing so, it raises larger issues of the relationship between technology and context, asking how ideas and circumstances affect action. As such, it is part of a growing body of work struggling to come to terms with this question, made more significant by increasing globalization and the growing impact of technology (computer-based or not) in our lives.

We believe that CSCW is a particularly appropriate object for this type of inquiry, since it is generally recognized as a field which spans a number of boundaries and integrates a variety of perspectives, ranging from those of hard science (engineering) to social science and even philosophy. As such, it can be thought of as a messy model or hybrid, in which the social and the technical are inextricably intertwined. The social "content" of a CSCW system is thus much greater that that of, say, a toaster or even a television. On the other hand, one cannot make abstraction of the very real technical knowledge and constraints that go into building a working system.

This paper suggests that CSCW systems, like all technologies, can be read as texts. These technological texts contain some elements that are distinctive to their culture of origin, without necessarily being unique to that context. It further offers a plausible explanation for these design choices, basing its argument in the discourse of designers themselves. It draws on the notion of technological frame (Bijker and Law 1992) to explain how Japanese CSCW designers invoke Japanese culture in general and certain aspects in particular as resources upon which to found technical decisions, illustrating the translation of these cultural arguments in CSCW systems.[1]

Background

Cultural attitudes towards technology and cultural dimensions in the implementation and use of technology are topics of increasing interest worldwide, perhaps as a result of increasing globalization and intercultural contact. This subject is becoming all the more significant with the proliferation of new communications technologies that hold out the promise of global communication. The novelty of new computer-mediated communication networks does not, however, mean that we must start from scratch in attempting to understand how people from different cultures will use them, and how diverse cultural attitudes are likely to affect their use. Over the past twenty years these questions have in fact been explored in the fields of both organizational and development communication.

In development communication, a turn-key approach to technology transfer has been rejected in favor of other models which accord substantial importance to culture. Among them, there has been considerable research on the importance of technological infrastructure and predisposition or competency as preconditions for technology transfer (Andrews and Miller 1987; Copeland 1986), as well as various measures for increasing the likelihood of successful transfer: modification of imported technology by local engineers to make it more "appropriate" (De Laet 1994; Ito 1986), a two-step flow in which new ideas or technology are introduced first to an opinion leader or technological gatekeeper who then persuades others to adopt it (Rogers and Shoemaker 1971), or involving stakeholders in planning and decisions (Ackoff 1981; Madu 1992). All this work shares a concern for facilitating accommodation to a changing environment produced with the introduction of new technology. In other words, making the technology fit its context of implementation and use has been found to considerably improve the chances of optimal use.

Understanding the reciprocal link between organizational practices and technologies has also been a key concern of organizational communication scholars, particularly with the advent of office automation and computerization. Many have drawn on Giddens' structuration work (Orlikowski and Gash 1994; Orlikowski 1992; Poole and DeSanctis 1990) to explain how computerization changes organizational structure. Heath and Luff (1994) have studied the evolution of social interaction in technological environments. In the field of information systems management, several authors have suggested that differences in national culture may explain differences in IS effects (Deans and Ricks 1991; Raman and Watson 1994; Watson and Brancheau 1991).

In short, studies in development and organizational communication over the past two decades have consistently pointed to three key factors in explaining successful IT implementation: existing technological infrastructure and predisposition—the context; the process of implementation; and the importance of viewing use as a process in which uses change over time. This is evidenced in needs and gratifications, and active reception theories of communication.

At the same time, there has been a growing backlash against technological determinism, an increasing awareness that the path a given technology takes may not be inevitable and absolute. Although many engineers may continue to support the position that the technologies they build are neutral, it has become something of a commonplace in the social sciences to say that technology is socially constructed. In recent years, numerous instances of how technical artifacts embody political, cultural or economic positions have been identified (see for example the collections edited by Bijker, Hughes and Pinch 1987 and Bijker and Law 1992, as well as Winner 1993). Increasingly, it appears important to understand how technological artifacts are constructed and how the end result relates to its conditions of construction if we are to understand their implementation and use.

The challenge for social science, in our view, is to go a step further to examine how this process of social construction is accomplished and to determine which aspects of the black box called "technology" are more or less susceptible to social influences. By asking how ideas and circumstance affect action, we are in fact raising larger issues of the relationship between technology and context. As such, this research is part of a growing body of work struggling to come to terms with this question of growing significance given increasing globalization and the increasing impact of technology (computer-based or not) in our lives (Hales 1994; Jackson 1996).

Research Question and Method

This paper focuses on one object: computer-supported cooperative work (CSCW), one stage in the process: design, and one cultural context: Japan. It is based on a larger, comparative study (Heaton 1997) whose central research question was the extent to which different preoccupations in different countries are the result of different "cultural constructions of computing." How do CSCW designers translate their ideas about what people do when they work, and the role of computers in supporting work, into the systems they design? What is the impact of the circumstances[2] in which designers find themselves, on the systems they design?

Given the complexity of the subject matter, and the small number of laboratories actually involved in CSCW design, we adopted a case study approach as an appropriate means of capturing the subtleties of the multitude of situational variables and their interaction. During five months of observation in various CSCW laboratories, the author conducted extensive interviews with over twenty software designers and took part in numerous informal conversations with others involved in CSCW research. Earlier typologies of cultures, particularly as they have been applied to the world of work, were used as a starting point and a general guide for observation, although no attempt was made to fit the data gathered into these classificatory schemes. Analysis of documents produced by the laboratories in question was also an important part of the process. Some of these documents described the CSCW systems, while others were explanatory in nature. Both internal (working documents, memos, project reports) and external documents (scientific publications) were analyzed. The focus was twofold: to understand how designers perceive their work through what they say and write about it, and to analyze the work itself (both work practices and the resulting machines and software), the goal being to draw parallels between the two.

The present paper focuses primarily on the relationship between designers' justifications for their choices and how these choices are reflected in the design of machines and software. The specific cases presented are illustrative of larger tendencies and trends in CSCW design in Japan.

Patterns in CSCW Research

In the context of this paper, CSCW has been broadly defined as work by multiple active subjects sharing a common object and supported by information technology. The presence of active subjects provides a means for delineating CSCW from traditional office automation perspectives. Furthermore, a community that shares a common object of work can always be delineated in practice, whatever the contributions of the different participants. The focus of computer supported cooperative work, then, is less on working with computers than on working with each other through computers. This changing orientation opens the door to a real contribution from social scientists to understanding the complex relationship between technology and its context of emergence and implementation.

A quick survey of the CSCW literature points to an amazing variety of "solutions" or approaches to similar problems. What is more, these solutions seem to follow certain patterns. Not only are there very real differences between the various communities of practice involved in CSCW[3], the field also demonstrates marked regional differences in emphasis and perspective. American CSCW has tended to take an empirical approach and to focus on product development and small-group applications, while Europeans are generally more theoretical or philosophical in orientation and tend to focus on the user organizations and organization systems. In Japan, considerations have generally been pragmatic and there is considerable interest in formal workflow management systems and the software factory concept.

A systematic review of the CSCW and European CSCW conference proceedings over the past decade (Heaton 1997) documents a number of general patterns in how CSCW researchers present their work to the international academic community of their peers. Presentations coming out of Japan illustrate a considerable homogeneity in research interests. All the research presented at international CSCW conferences has centered on the exploration of the possibilities of video, multimedia, and large screen displays. Gesture has a major importance, as does shared view of workspaces. Japanese work tends to present solutions which are technically innovative and which require major investments of technical resources (high bandwidth communication channels, large flat screen displays, a number of video cameras, etc.) Finally, the Japanese groupware scene is much more technically oriented than European or American contexts. Japanese researchers readily admit to their technical focus and product orientation. In fact, one of the prime criteria for evaluating a research project appears to be whether or not it is up and running, and it is inconceivable for the researchers interviewed that research not lead to a working system.

In contrast, video-mediated communication is completely absent in Scandinavian work, which focuses on organizational issues and is typically presented in the form of cases in which designers have been active participants. Cooperative design, supporting users in their daily work, and looking at work as situated in a specific context are common themes. British work is fairly equally distributed among case studies, conceptual and technical articles, while the volume and variety of work done in North America makes it very difficult to classify; all tendencies are represented, from high-tech video-intensive environments, to ethnographic studies of implementation and use, to theoretical models of coordination.

The question remains: how can we explain that designers, who have similar technical knowledge and professional backgrounds, choose to explore different issues or questions, and, what is more, appear to answer them in different ways? This is all the more astonishing given the fact that they identify themselves as members of the same research community and are in regular contact with designers from various countries and institutions. Clearly differences between communities of practice alone cannot explain these differences in orientation. Grudin (1991a, 1991b) has outlined a number of partial explanations including institutional support, funding, even cultural norms; others have applied an actor-network approach to analyze the political and cultural regimes in which design is embedded in specific cases (Gärtner and Wagner 1994; Hakken 1994). Here, we seek an explanation for regional differences in CSCW not in institutional variables, nor in strictly professional ones, but at a mid-level between micro and macro—in culture, which is both an individual attribute and a collective phenomenon. Field research provides concrete illustrations of the importance of culture as a variable in the technology design process.

On Culture

While Japanese CSCW design is the focus of this paper, this should not be taken to imply simply a discussion of national culture. As will become clear in the discussion of our cases, organizational and professional cultures are also vital elements in the mix. First, however, some background and clarification of what we mean by culture is in order.

The movement to distinguish between national cultures finds its roots in social anthropology of the thirties and forties. More recently, forces in the real world have heightened awareness of the importance of the cultural factor and a number of studies on work organization and work attitudes have consistently demonstrated significant differences across national cultures. Among a number of typologies of cultures, the most widely cited and one of the most thorough is that of Geert Hofstede. In an attempt to identify cultural predispositions that Bourdieu has called *habitus*[4], Hofstede (1980) administered standardized questionnaires to some 116,000 people working for IBM in a variety of professions in over fifty countries in 1968 and again in 1972. On the basis of this data, Hofstede defined several dimensions of culture.[5] This and other similar studies clearly indicate that people from different cultures bring different attitudes to their work and that this results in national differences in the way work is organized and work practices.

Japan, for example, can be characterized as a group-oriented society with a long-term orientation, strong uncertainty avoidance, and highly differentiated gender roles, and which accepts the unequal distribution of power. North American society, on the other hand, is highly individualistic and less tolerant of the unequal distribution of power, with a short-term orientation, and medium degrees of uncertainty avoidance and gender role distinction. The four Scandinavian countries form a relatively homogeneous group, with few gender distinctions and generally low power distance, more group-oriented than North America but less so than Japan.

Another body of literature has examined differences in attitudes, values and practices between professions. A person's occupation or training undoubtedly has a major influence on how he or she approaches the world. For example, computer scientists likely draw on a similar pool of knowledge and techniques relative to systems development, which in turn calls for and constitutes a particular way of looking at the world.[6] Similarly, social scientists may not always share common frames of reference but most will share certain elements of common knowledge. In the case of CSCW, it is probably justifiable to distinguish a third general professional group, composed of managers and end-users.

Professional culture becomes a central concern as soon as communication between communities of practice becomes necessary. Systems engineers may be operating from one set of assumptions, while those studying the work practices the system is designed to support or supplant may have a fundamentally different perception of the task at hand, and those who initiated the project (upper management, unions, etc.) yet other objectives and perceptions. The negotiation of shared meanings is a key research issue in CSCW.[7]

Ulf Hannerz (1992, 249) has coined the term transnational cultures, which he defines as "structures of meaning carried by social networks which are not wholly based in any single territory." Many transnational cultures are occupational. Hannerz suggests that, while it makes sense to see them as a particular phenomenon, they must at the same time be seen in their relationships to territorially-based cultures and argues that their real significance lies in their mediating possibilities. While "transnational cultures are penetrable to various degrees by the local meanings carried in settings and by participants in particular situations" (251), they also provide points of contact between different territorial cultures.

The important point here is that occupational culture need not be a subset of national culture. Rather, the two are distinct and interrelated. Those involved in CSCW system design share a common "CSCW culture"[8], but they also reflect and interpret this professional culture within the framework of their territorial cultures, just as professional training and perspectives lead them to interpret elements of territorial culture in certain ways. A given situation, say the design of a particular CSCW system, can be understood in cultural terms as the product of what is unique (national culture) and what is shared by all (occupational culture). The resulting combination of the two will necessarily differ between cultures and even between systems in the same national culture, because conditions can never be identical.

Finally, there is organizational culture, which is perhaps best understood as a root metaphor. Starting with the premise that organization rests in shared systems of meaning, and hence in the shared interpretative schemes that create and recreate that meaning, it directs attention to the symbolic or even "magical" significance of even the most rational aspects of organizational life and calls for recognition of the complexity of everyday (organizational) life. Erez and Earley (1993, 69) cite a number of empirical studies that suggest that national or societal culture must be considered along with organizational culture in order to fully understand the relation of an organization's culture to its functioning.

In summary, for the purposes of this research culture is defined as a dynamic mix of national/geographic, organizational, and professional or disciplinary variables in constant interaction with one another. Culture changes according to context and over time, and should be understood not in terms of pre-existing, fixed categories, but as resources, accumulations of actions, patterns that constitute, reinforce and transform social life. In short, culture is continually constructed and reconstructed.

Culture in the Frames of CSCW Researchers

The notion of technological frame provides an interesting way of approaching culture from a constructivist perspective. Law and Bijker (1992, 301) use the notion to "refer to the concepts, techniques and resources used in a community—any community. . . . It is thus a combination of explicit theory, tacit knowledge, general engineering practice, cultural values, prescribed testing procedures, devices, material networks, and systems used in a community." It is simultaneously technical and social, intrinsically heterogeneous. The related expression 'frame of meaning' as coined by Collins and Pinch (1982) and adopted by Carlson (1992) in his study of Edison and the development of motion pictures, translates the specific focus of this paper on how cultural patterns and assumptions inform actions and shape choices most closely:

> . . . in any given culture there are many ways in which a technology may be successfully used . . . To select from among these alternatives, individuals must make assumptions about who will use a technology and the meanings users might assign to it. These assumptions constitute a frame of meaning inventors and entrepreneurs use to guide their efforts at designing, manufacturing, and marketing their technological artifacts. Such frames thus directly link the inventor's unique artifact with larger social or cultural values. (Carlson 1992, 177)

Carlson argues that designers attempt to impose pre-existing frames based on previous experience on new products or invention, rather than inventing new frames. This unconscious process of "cultural creep" results because designers create artifacts to fit into the cultural spaces suggested by their existing frames of meaning. It is only after their introduction that new uses and new cultural meanings are developed. Thus, users are present virtually in designers' frames, whether or not an artifact has actually been used (Flichy 1995). The distinction between design and use thus appears more of an analytic convenience than a hard and fast rule. Consequently, we suggest that it may be more valuable to approach design-implementation-use as a single process, in which all stages are interrelated.

The following section presents the world of two Japanese CSCW laboratories, with a view to highlighting common research themes. A brief description of the overall context of CSCW design in Japan is followed by detailed presentation of two research projects. The section concludes with a discussion of general trends and characteristics and relates them to cultural characteristics and beliefs, which are intimately connected to designers' views of their systems' eventual use.

Japanese CSCW: Quality (and Quantity) of Work

CSCW in Japan is a development of the telecommunications, electronics and engineering industries and is thus closely identified with a product, rather than a research orientation. A "hard" science approach dominates. Virtually all those involved in designing CSCW systems in Japan are engineers or computer scientists. They identify strongly with their profession, and building a good system, that is one that works, is reliable, state-of-the art, and original, is both the goal and a measure of their capabilities as engineers. Design work is done exclusively in the labs, and any evaluation of prototypes takes the form of controlled laboratory experiments. Designers are not generally concerned with who will use their systems, or how they will be implemented. Multidisciplinary collaboration is not considered, let alone practiced.

With so technical a focus, it is not surprising that the main justifications for design choices are technical ones. There is however, another, more social, element to Japanese design choices, that of Japanese culture. Professional engineering or scientific culture notwithstanding, Japanese CSCW researchers, like most Japanese people, clearly believe that Japanese culture and the Japanese way of working are different from the Western ways.[9] How to reflect or cope with this difference in designing technology is a constant *leitmotif* among Japanese CSCW researchers. Although most would prefer to believe that science and technology are culturally neutral or universal, they nevertheless recognize that, if use is a consideration, designing a groupware system cannot be approached the same way as designing a television.

The dean of groupware in Japan, Professor Matsushita, cites five principal, specifically cultural reasons why groupware must be different if it is to be used in Japan: cultural differences in views on cooperation and competition, negotiation style, degree of context, the importance of human relations, and the relation of the individual to the group. Even those who deny specifically cultural aspects in the design of CSCW and groupware in Japan, acknowledge cultural effects in implementation and use. Some major Japanese companies are now selling workflow systems developed by American companies, but this is problematic. In the words of another leading researcher, the biggest challenge facing Japanese groupware is "attaining widespread use. Managers don't want to change the way they work. They want to be able to consult with people as they usually do."

How does this desire to reflect cultural particularities play out in practice?

TeamWorkStation / Clearboard (NTT HUMAN INTERFACE Labs)

Our first example, TeamWorkStation, is one of the earliest and most documented Japanese CSCW projects. It has been widely cited within the CSCW community and has inspired considerable research within Japan around the concepts of seamlessness and gaze awareness. Ishii and his collaborators at NTT Human Interface Labs were neither the first to develop the concept of a seamless work environment, nor the first to explore peripheral awareness. Both were borrowed from work done originally at Xerox PARC. But the Japanese way of dealing with these issues is unique, and the progression from TeamWorkStation I to TWS II to ClearFace to ClearBoard is illustrative of incremental development of research intuitions as well the resolution of technical problems.

TeamWorkStation (TWS) is "a desktop real-time shared workspace" which integrates both computer and desktop workspaces. Starting from the premise that "no new piece of technology should block the potential use of already existing tools and methods" (Ishii and Miyake 1991, 39), the team set out to design a system that would allow users to maintain their preferred work practices, using their preferred computer applications, or even working with pencil and paper within a shared virtual workspace. Acknowledging that people might not do everything by computer and supporting their continued use of paper-based media were revolutionary concepts in CSCW at the time.

A second design requirement was a shared drawing surface. The research team chose video as the basic media of TWS for its ability to fuse traditionally incompatible media such as papers and computer files (Ishii and Miyake 1991, 39). Live video image synthesis was employed to capture individual workspaces (both computer screens and physical desktops) and to display them in separate layers on a computer monitor. The overlay function created with this technique allowed users to combine individual workspaces, and to point to and draw on the overlaid images simultaneously.

The three-member design team began to use the prototype on a daily basis in July 1989, and informal evaluations of its use pointed to the importance of gesture as a means of enforcing the sense of shared space. They preferred hand gestures to pointing or marking with a mouse "because hand gestures are much more expressive, and because hand marking is generally quicker" (Ishii and Miyake 1991, 45). Since the TWS prototype was designed without a formal floor control mechanism for passing the input control among collaborators, voice contact played an important role in preserving informal social protocol and coordinating action, especially the use of the limited workspace on the shared screen (Ishii and Miyake 1991, 45).

The faces of collaborators were displayed in separate windows beside the shared workspace in TWS. But spatial awareness was already a concern, and was developed further by ClearFace and later ClearBoard. All previous approaches to CSCW screen layout (tiling—i.e. laying them side by side—or overlapping windows) required users to shift their focus between the shared drawing space and the facial images and deal with separately. Developed initially as a solution to a technical problem, how to make the most of limited screen size (14" in the TWS prototype), the ClearFace interface proposed translucent, movable, and resizable face windows which overlay the shared workspace window. The user could see the drawing space and his collaborators' faces in the same space and shift easily between the two. The team explained this facility using Neisser's theory of selective looking and the high recognizability of human features, further reasoning that it is rarely necessary to attend to both at the same time (figure ground relationship), thus eliminating possible confusion of different "layers." In use, they observed that people hesitated to draw or write over people's faces, inciting them to make the face windows movable and resizable.

With ClearFace, the design team began to explore the dynamic relationship between elements in design meetings. Their focus shifted away from task—what workers are doing—to how they are relating to each other as they do it. In one of their later papers, Ishii et al. present this change as a transition from a focus on shared workspaces to the creation of interpersonal spaces (Ishii, Kobayashi and Grudin 1992, 33).

At the same time, in the discussion, the participants are speaking to and seeing each other, and using facial expressions and gestures to communicate. In the conversations it is essential to see the partner's face and body. The facial expressions and gestures provide a variety of non-verbal cues that are essential in human communication. The focus of a design session changes dramatically. When we discuss abstract concepts or design philosophy, we often see each other's face. When we discuss concrete system architectures, we intensively use a whiteboard by drawing diagrams on it (Ishii and Arita 1991, 165).

The effort to simulate as closely as possible the collaboration in front of a whiteboard was taken a step further in ClearBoard, the first prototype to refer explicitly to eye contact and gaze awareness (see Figure 1). The design metaphor here was talking through and drawing on a transparent glass window. The system used colored markers on a glass board, and video and a half-mirror technique to capture and orient the drawings. In this case, users recognized their partner as being behind a glass board and they did not hesitate to draw over the facial image. The large size of the drawing board supported awareness of gesture and of the partner's surrounding environment, as well as of his visual focus.

The most novel feature of ClearBoard, and the most important, is that it provides precise "gaze awareness" or "gaze tracking." A ClearBoard user can easily recognize what the partner is gazing at on the screen during a conversation. The importance of eye contact is often discussed in the design of face-to-face communication tools. However, we believe the concept of gaze awareness is more generalized and is a more important notion. Gaze awareness lets a user know what the partner is looking at, the user's face or anything else on the shared workspace. If the partner is looking at you, you know it. If the partner is gazing at an object in the shared workspace, you can know what the object is. Eye contact can be seen as just a special case of gaze awareness (Ishii and Kobayashi 1992, 530–531).

Gaze awareness allows participants to better situate the interaction within its context, providing a wider variety of cues for feedback and a richer awareness of the environment and others' activities. The emphasis on non-verbal cues and direction of gaze rather than eye contact is particularly significant coming from a culture in which eye contact is much less common than in Western culture and is in many cases considered rude. Indeed, Ishii et al. make a veiled reference to this problem: "ClearBoard makes eye contact easy to establish and may even make it more difficult to avoid. It has been shown that the use of eye contact varies with the culture (e.g. Argyle 1975); these are issues for further exploration in ClearBoard settings" (Ishii, Kobayashi and Grudin 1993, 372).

Figure 1
Clearboard

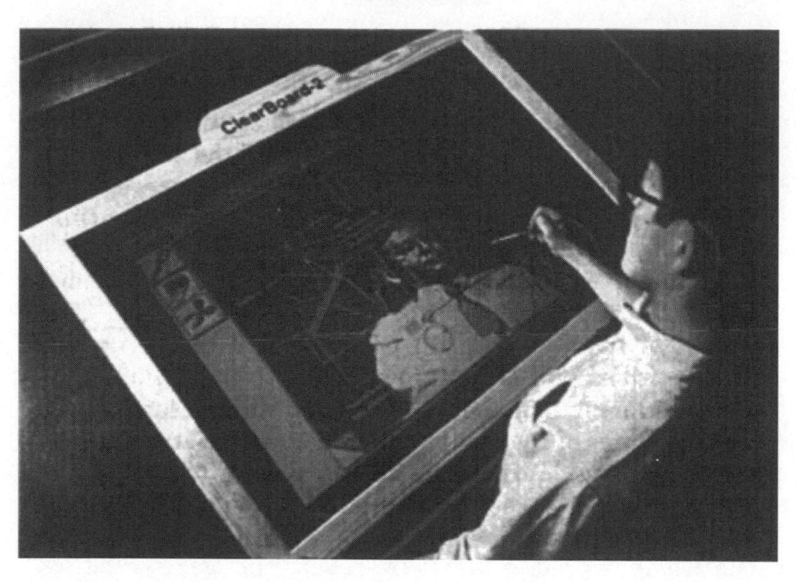

©ACM, 1993, *TOIS*, 11 (4) Ishii, Kobayashi and Grudin

Several technical problems present in ClearBoard-1 (low video resolution, forcing the use of thick markers which quickly used up the drawing space, and the inability to alter the partner's drawing in shared video drawing) were resolved in ClearBoard-2, an extension of the same idea but this time using computers. Multi-user drawing software and digitizer pens were used to permit the direct recording of work by any number of users simultaneously. This also allowed the integration of computer files into the system, and enabled the results of design sessions to be saved as PICT files. Finally, the ClearBoard-2 design led to some reflections on interpersonal distance:

> ClearBoard creates the impression of participants standing about *one meter apart*, because both sit (or stand) close enough to the screen to draw directly on its surface. This virtual distance belongs to the *personal distance* in Hall's classification. When people use ClearBoard with close friends or colleagues, this distance seems appropriate. However, for a formal meeting with a person of much higher rank, this virtual interpersonal distance might seem too small, and the

> participants might be uncomfortable. Therefore, we would like the media to provide users with some control over the virtual interpersonal distance. We are planning to provide an option of indirect drawing using a wireless tablet or pen-based personal computer for that purpose. (Ishii, Kobayashi and Grudin 1993, 371–72)

While the NTT Human Interface Labs team was disbanded before they were able to pursue this research direction, the concern for interpersonal distance was picked up and further explored by another research group in our next case, MAJIC.

MAJIC (Matsushita Lab, Keio University)

Our second case is a system developed at the Matsushita Lab in the Instrumentation and Engineering Department of Keio University, a prestigious private university located near Tokyo. MAJIC illustrates many research themes characteristic of Japanese CSCW. To a large extent, it builds on earlier Japanese work at NTT on eye contact and gaze awareness, adding a multiparticipant dimension and a more explicit focus on the surrounding environment. This relationship to earlier work is both professional and personal. In addition to the bibliographic citations in published papers, one of the designers told me specifically that he was influenced by Dr. Ishii's work on gaze awareness. Furthermore, one of the Clearboard designers was his *sempai* (upperclassman) at Keio University. The MAJIC team explains clearly why they feel this line of inquiry is important:

> When we have discussions in face-to-face situations and people approve of a statement, we can tell by their attitude, tone, eye movements, gestures and so forth, whether or not they approve wholeheartedly. It is difficult, on the other hand, to estimate how strongly they approve when we read only the minutes without attending a meeting. Hence, one of the purposes and/or advantages of face-to-face meetings is that all of the participants are *aware of the speaker's intent and the other listeners' reactions* based on both verbal and nonverbal communication. (Okada et al. 1994, 385)

As in TeamWorkStation, there are multiple references to the importance of context, orientation to the other (how what you say is being received), and a focus on interpretation of intention rather than surface meaning. The key design issues of MAJIC were defined as support of multi-way round-table meetings and multiple eye contact; maintenance of peripheral gaze awareness; seamless presentation of life-size images of participants to achieve a sense of reality; and a shared work space (Okada et al. 1994, 385).

The creation of a seamless environment and sense of presence in MAJIC relies extensively on non-verbal behavioral information, such as eye contact, gaze awareness, gesture and body language, and on contextual cues such as image size, distance and background. References to these elements are extremely specific. For example, the MAJIC team refers to symmetrical or asymmetrical postures and body orientations as important cues: "In this way we sense the atmosphere in the meeting room and the aura of the participants, and, consequently, we can understand the opinions of the participants clearly and make the meeting productive" (Okada et al. 1994, 386). They cite gaze as a means of controlling a meeting: "A chairperson sometimes gazes at participants to urge them to speak when there is silence in a meeting" (386). They also discuss the social uses of eye contact: "Of course eye contact is very important in communicating with one another, as mentioned above, but especially in Japan it is impolite to look into someone's eyes for a long time" (387). In their observations of face-to-face meetings, the designers noted that participants most commonly averted their eyes by looking down at material on a table in front of them, and decided to provide such a table in their design (390).

Referring to Hall's (1976) classifications of appropriate distances for interactions, the MAJIC team discusses elements which may affect virtual distance (the sensed distance among participants): physical distance from the display, the size and quality of video images, voice fidelity, backdrop, etc. In fact, this has been the central focus of most of the MAJIC research. Starting with the assumption that image size of participants and background are the two important factors in achieving a sense of reality during videoconferencing, MAJIC I was designed to project life-size video images and to simulate a virtual social distance of approximately four feet between participants.

The central element of MAJIC is a large (four-by-eight feet), curved semi-transparent screen. Each MAJIC unit also contains a workstation (with a recessed, tilted monitor), two video projectors, two video cameras, two directional microphones and two loudspeakers. Video images of the participants are projected onto the screen and captured from behind it. Each participant sees the frontal view of the others and the edges of the images overlap slightly (see Figures 2 and 3).

Figure 2
First draft of MAJIC

The second factor deemed essential for "achieving a feeling of togetherness during videoconferencing" (390) is the continuity of background images. In this interpretation of "seamlessness" if images run into each other, it is difficult to tell where one ends and the next begins; "if users are surrounded by other participants with a seamless background, they can feel as though they are together." (386) In actual fact, the backgrounds must be "matched" at the seam. But this is only a prototype; MAJIC proposes doing away with the actual background altogether and replacing it with an artificial one that can be chosen to create a desired mood, to relax or to inspire (386 and personal communication). This would be done using a chromakey blue background.

Laboratory evaluations of MAJIC I and observations of use at a trade fair led to several improvements or additions in MAJIC II. For example, in a questionnaire administered to forty students, three-quarter size images were rated more convincing than life-sized ones. This led to experiments to determine the ideal relationship between distance from the image and image size and an adjustment in MAJIC II. There have also been a number of strictly technical improvements: improvement of image quality, reduced size of the prototype, etc.

Figure 3
Gaze awareness in MAJIC

A further extension of the idea of direct physical manipulation in MAJIC II is the "Whisper Chair." By leaning right or left, the person sitting in this chair (equipped with sensors) can talk to one or the other persons on screen without the third party hearing. The rationale behind this development is that leaning is a more subtle, more natural way of confiding a secret than flipping a switch to turn the audio channel off.

MAJIC represents a curious mix of virtual or imaginary space and an interest in simulating reality as closely as possible, including providing direct physical feedback whenever possible. In the demonstration video of MAJIC shown at CSCW'94, the participants have a "virtual tea party" in which one person "pours" and real tea comes out into the cup of another. Although this is a presentation gimmick (and the metaphor of sharing tea is highly significant in Japanese culture), one is left wondering where the limits might be.[10]

Discussion

Characterizations of Japan as a society in which human relations are all-important, relationships are dependent on positioning people on vertical (hierarchy) and horizontal (in or out-group) axes, and where communication is highly indexical or context-dependent, have been widely discussed in the business and sociological literature on Japan [see for example Stewart (1987), Ito (1989) and Barnlund (1989) specifically on interpersonal communication in organizations]. The extent of agreement in the literature suggests that they are firmly grounded in reality.

Edward T. Hall (1976), an author cited by CSCW researchers, uses the terms high- or low-context culture to refer to a culture's preferred communication style: the degree to which the meaning of a message can be abstracted from the situation in which it was produced and received. A high-context message is one in which "most of the information is either in the physical context or internalized in the person, while very little is in the coded explicit transmitted part of the message"; a low-context message is one in which "the mass of information is vested in the explicit code" (Hall, 1976: 91). The concept has implications for implicit/explicit, verbal/non-verbal, affective or intuitive/ fact-based, and relational/absolute communication. In a society like Japan where most behavior and the use of language is highly codified, the form is standard. It is important to look beneath the surface to interpret the meaning of an exchange, hence the importance of positioning and the emphasis on atmosphere. Much of the content of a message will be implicit; interpretation will often be based on intuition rather than facts; and relationships will continually shift and be redefined.

Several common traits emerge in Japanese designers' attempts to deal with the particularities of their culture. First, fully conscious of the highly relativistic approach to relationships in their society, designers do not believe that all types of communication can be supported by groupware systems. All readily admit that there are limits to supporting the more subtle or situationally-dependent aspects of work. Given the constantly fluctuations and redefinitions involved in any activity which is out of the ordinary, they view the task of trying to support "delicate" communication, such as negotiation, as an impossible one. One researcher points to the impossibility of "catching" pieces of information which fly around an office and are grasped through peripheral awareness. Despite listing a shared workspace as one of the design issues and providing a workstation and table, no one has yet tried to work using MAJIC, even in the laboratory. And the NTT Software Labs team's research shifted in focus from shared workspace to interpersonal interaction during work.

A corollary of not trusting a computer system to model all instances of human communication or to successfully translate the subtleties of day-to-day interaction, is the focus of many Japanese CSCW systems on providing channels for communication rather than trying to specify content or process. By providing a variety of channels, nothing more, nothing less, a CSCW system should ideally be able to support all kinds of communication regardless of the message content or objective. This is clearly the case with MAJIC in which research and evaluation have focused exclusively on the physical environment. In TeamWorkStation/Clearboard, too, the focus is on providing an environment which simulates as closely as possible a face-to-face situation and which does not in any way constrain potential use.

Another feature of Japanese CSCW systems is that they are careful to provide support for traditional, paper-based forms of working, and ways of integrating paper and electronic information. Designers view the systems they design as complementary to, not replacements for standard practices; their aim is to support groups, not to replace or reconfigure all their activities. TWS and Clearboard use video to capture texts or drawings on paper. The MAJIC system integrates a desk that people can work on. These systems also allow people to draw using pen or pen-based computing technology. This is all the more significant considering the transformations involved in converting keyboard input to Japanese ideograms or *kanji*. As one informant notes, "typing is not easy for us."

When language cannot convey all meaning, nonverbal communication becomes more important. Perhaps most significantly, Japanese CSCW systems are also characterized by extensive emphasis on providing contextual cues so that Japanese using these systems will be able to orient their behavior appropriately. This emphasis on the contextual translates into research on spatial awareness, gaze awareness rather than eye contact, gesture, interpersonal distance, physical feedback, and large displays. One informant even went so far as to insist that physical feedback must be integrated into the interface design because he does not believe it is possible for Japanese to have an entirely intellectual relationship with the computer.

Furthermore, considerable attention is paid to creating a pleasurable physical environment or a shared environment, as in TWS or MAJIC, with tones of virtual reality. If a CSCW system is to be useful in Japan, it is important that a sense of atmosphere or feeling transpire through the system. A Japanese psychologist whose research interest is group dynamics tells me that the most important thing in Japanese groups is face-to-face communication, which creates atmosphere, or *kuuki*.[11] This is borne out by use experiments of several CSCW systems which have demonstrated that it is difficult for a group to use them without having first met to establish an atmosphere of mutual trust. "We need to meet once face-to-face before having such a meeting because without meeting face-to-face we don't feel friendly or we don't feel easy to talk. . . . And once we have met we can use such kind of machine. But we thought we still need video images to make the participants feel easy or feel friendly."[12]

The cases presented above illustrate the close relationship between designers' preconceptions and frames of reference and the systems they design. Japanese CSCW researchers consistently invoke Japanese culture as a justification for decisions to focus on contextual awareness and non-verbal communication in Japanese CSCW systems. The preferred Japanese approach to CSCW design is to provide a channel for communication, which can be used to complement, or supplement, traditional ways of working. This channel should transmit as much information as possible (hence the widespread use of video and large displays) but should avoid specifying procedures or ways of doing things. It is not a tool, but another element in the working environment that can offer important contextual information to enable coworkers to evaluate a situation and to respond in accordance with existing social protocols.

While certain characteristics of Japanese CSCW systems can be explained with reference to the particularities of their society, it is also significant, in our view, that there is such widespread agreement on what constitutes interesting CSCW research in Japan. Ishii's work on gaze awareness and the use of video have been picked up and pursued by the Japanese CSCW community. Similarly, the importance of gesture, body language and postures in supporting awareness between coworkers, and considerations of interpersonal distance are recurrent themes. Certainly, these issues must strike a chord as designers try to build systems that will correspond to potential uses and eventual contexts of use as they understand them.[13]

Implications

Clearly, the frames of meaning of Japanese CSCW researchers have a major impact on their design choices. These choices in turn guide the implementation and eventual use of these systems. Designers create artifacts to fit into cultural spaces as they understand them.

New uses and new cultural meanings can only be developed after the fact. It is too early to tell whether or not CSCW designers are justified in their attention to non-verbal, contextual support. Japanese CSCW has been criticized for simply trying to simulate face-to-face reality as closely as possible and for neglecting to exploit some of the transformative potential of computer mediated communication. We would like to suggest that, rather than abdicating responsibility for the consequences of their designs, Japanese designers have adopted a pragmatic approach: designing for use as they understand it now, and leaving these uses to develop as they will.

The explicit cultural sensitivity of Japanese CSCW work also point to a need for cultural sensitivity in the design of technological artifacts, and at a level that goes beyond ergonomics or changing surface details on an interface. In the case of Japan, the need for contextual information suggests that the use of language-based environments, even in Japanese, may be problematic. This difficulty goes far beyond the physical difficulty of inputting on a keyboard (although this is also a definite concern, as reflected in the extensive research on pen-based computing, speech synthesis and multimodal interfaces in Japan). There appears to be a demand for virtual reality interfaces, and initial experiments have demonstrated that VR-based interfaces to applications such as internet relay chat (IRC) are indeed very popular. Secondly, the assumed difficulty of fitting into a framework, or set way of doing things, suggests that organizing cooperative work as a series of procedures to be followed or channels to be taken may be inappropriate in Japan. In fact, this is confirmed by the choice of Japan's leading workflow expert to focus on the use of resources rather than the paths they follow.

We are only beginning to appreciate the complexity of the relationship between technology and its context and how changes in one inevitably affect the other. It is important to remember that technological artifacts are being designed by someone and that there is nothing inevitable about how they turn out. Design choices circumscribe a field of potential uses: some are built in, others are proscribed. Consequently, it is essential to consider design in studies of the implementation and use of technology.

Conclusion

This paper has outlined how designers' views on Japanese culture find their way into the design rationale for CSCW systems: Japanese

CSCW designers generally agree that Japan is unique and that designing for a Japanese context requires particular attention to a certain number of elements. Although it is not the only consideration in design, this attention to culture goes far beyond the stage of ideas to finds expression in the machinic reality of the computer systems, as illustrated by our two examples. The paper further proposes an explanation, grounded in the notion of cultural frame, for these observations. This explanation focuses on the interaction between the specific situation in which design is taking place, its larger social, cultural and institutional context, and the unique actions of designers. Based on how they understand the world around them, designers make assumptions that guide their design choices.[14] As participants in their larger professional, organizational and national cultures, individual designers link their creations with larger social or cultural values. They actualize their shared understandings of Japanese culture as they perform it in their daily design activities.

Acknowledgements

We would like to thank the Fonds pour la Formation de Chercheurs et l'Aide à la Recherche (FCAR), whose financial support enabled this research. Thanks are also due to all those who took the time to speak with us, particularly at NTT and Keio University.

This chapter appeared originally in the *Electronic Journal of Communication / La revue electronique de communication*, 8(3 & 4), 1998 (see <http://www.cios.org/www/ejcrec2.htm>), and in *AI and Society* (1999) 13: 357–376, and is reprinted by kind permission of the editors and publishers.

The illustrations in Figures 1, 2, and 3 are © by the Association for Computing Machinery, and are reprinted here by kind permission.

Notes

1. It is not the intention of this paper to demonstrate causality, and the author is well aware of the dangers involved in the retrospective reconstruction of intentions and influences from a finished product so characteristic of early SCOT (social construction of technology) work. It should simply be read within the larger objective of clarifying the relationship between what designers do and how they do it, and between what they do and what they say.

2. Circumstances here include the larger institutional context, as well as daily work practices, which serve as both resources and constraints on what can be done. While they provide structure, these resources and constraints should not be taken to be immutable.

3. The pervasive tension between designer/engineers on the one hand and social scientists on the other has been referred to within the CSCW world as the "great divide." It is increasingly recognized as a fact (even a defining characteristic) within the field (see Bannon and Schmidt 1991).

4. Bourdieu's idea is that certain conditions of existence produce a *habitus*, a system of permanent and transferable dispositions. A *habitus* functions as the basis for practices and images that can be collectively orchestrated without an actual conductor. [Editor's note: Sunny Yoon explores this notion more fully as a frame for her analysis of computer use in Korea, this volume.]

5. The first dimension, that of power distance, refers not the actual distribution of power, but to the extent to which the less powerful members of institutions and organizations within a country expect and accept that power is distributed unequally. This dimension has implications for hierarchy, centralization, privilege and status symbols. The individualism/collectivism dimension identifies the strength of ties to and belonging in a group. One might expect this dimension to be correlated with loyalty, trust, shared resources, even the relative importance of verbal or nonverbal communication. The masculinity/femininity dimension measures the clarity of gender role distinction, with masculine cultures having clearly defined gender, and feminine cultures considerable overlap. Finally, the uncertainty avoidance dimension measures the tolerance (or intolerance) of ambiguity, the way in which people cope with uncertain or unknown situations. In the workplace, one might expect correlations with the way the environment is structured, rules, precision and punctuality, tolerance of new ideas, as well as with motivation (achievement, security, esteem, belonging).

6. Although the training of computer scientists in Scandinavia, Japan and North America may also differ significantly in terms of "peripheral" components, with consequent implications for how they see their role. See Dahlbom and Mathiassen 1993 for a detailed description of the mechanistic, rational worldviews implicit in computer science and systems development.

7. The notions of communities of practice, boundary practices and boundary objects have been explored by a number of authors, including Brown and Duguid (1991, 1994), Wenger (1990), Star and Griesemer (1989).

8. This should not be taken to suggest that there one could identify a single CSCW culture. Far from it! It is surely more appropriate to talk about a mix of CSCW influences.

9. Mouer and Sugimoto 1986 trace the long history of the theme of Japanese uniqueness and suggest that, while the ideology of Japanese uniqueness has been used in the service of many interests, the basic assumption that all Japanese possess a common set of attitudes and share similar behavior patterns has remained largely unquestioned, particularly in English language publications. They conclude that the relationship between this ideology and views of Japanese society is maintained by a complex network of interpersonal and inter-institutional relationships. In other words, Japanology is a self-fulfilling prophesy, a social construction almost universally subscribed to.

10. In fact, the Matsushita Lab has continued to pursue its research into the blurring of the physical and the virtual. A recent presentation at the 10th annual symposium on User Interface Software and Technology (UIST) in Banff, Alberta (October 14–17, 1997) was entitled "A virtual office environment based on a shared room realizing awareness space and transmitting awareness information."

11. Maiya, personal communication 8–6–95. Maiya's interest in groupware is how kuuki might be transmitted at a distance.

12. Watabe, personal communication 23–6–95.

13. To some extent, Japanese researchers may also have been focusing on developing a distinctive Japanese style and building a reputation in the international community by choosing to emphasize the commonalities in their work.

14. Design choices are, of course, subject to constraint and enablement by situational variables which are actualized in a chain of events in the design process. How these come about would be the subject of another paper.

References

Ackoff, R. 1981. *Creating the Corporate Future*. New York: Wiley.

Andrews, S. B., and H. G. Milller. 1987 "Expanding Market Share: The Role of American Corporations in Technical Assistance." *International Journal of Manpower* 6: 25–27.

Bannon, L. and K. Schmidt, K. 1991. "CSCW: Four Characters in Search of a Context." In *Studies in Computer Supported Cooperative Work*, eds. J. Bowers and S. Benford, 3–16. Amsterdam: North-Holland.

Bansler, J. 1989. "Systems Development in Scandinavia: Three Theoretical Schools," *Office, Technology and People* 4: 117–133.

Bijker, W. E., T. P. Hughes, and T. J. Pinch, eds. 1987. *The Social Construction of Technological Systems: New Directions in the Sociology and History of Technology*, Cambridge, Mass.: MIT Press.

Bijker, W. E., and J. Law, eds. 1992. *Shaping Technology/Building Society: Studies in Sociotechnical Change*. Cambridge, Mass.: MIT Press.

Bourdieu, P. 1983. *Le métier de sociologue: préalables épistémologiques*. 4th edition. Berlin: Mouton.

Brown, J. S., and P. Duguid. 1991. "Organizational Learning and Communities of Practice: Toward a Unified View of Working, Learning and Innovation." *Organizational Science* 2(1): 40–57.

———. 1994. "Borderline Issues: Social and Material Aspects of Design." *Human Computer Interaction* 9 (1): 3–36.

Carlson, W. B. 1992. "Artifacts and Frames of Meaning: The Cultural Construction of Motion Pictures." In *Shaping Technology, Building Society*, eds. W. Bijker and J. Law, 175–98. Cambridge, Mass.: MIT Press.

Collins, H., and T. Pinch, T. 1982. *Frames of Meaning: The Social Construction of Extraordinary Science*. Boston: Routledge and Kegan Paul.

Dahlbom, B., and L. Mathiasson. 1993. *Computers in Context: The Philosophy and Practice of Systems Design*. Oxford: NCC Blackwell.

Deans, C. P., and D. A. Ricks. 1991. "MIS Research: A Model for Incorporating the International Dimension." *Journal of High Technology Management Review* 2 (1): 57–81.

DeLaet, C. 1992. *Des Outils pour un développement durable: plaidoyer pour une réconfiguration*. Ph.D. diss., Université de Montréal, Montréal.

Erez, M., and P. C. Earley. 1993. *Culture, Self-Identity, and Work*. New York: Oxford University Press.

Flichy, P. 1995. *L'innovation technique*. Paris: La Découverte.

Gärtner, J., and I. Wagner. 1994. "Systems as Intermediaries: Political Frameworks of Design and Participation." *Proceedings of the Participatory Design Conference 1994*, 37–46. Palo Alto, CA: CPSR.

Grudin, J. 1991a. "Interactive Systems: Bridging the Gaps Between Developers and Users." *Computer* 24 (4): 59–69.

———. 1991b. "Obstacles to User Involvement in Software Product Development, With Implications for CSCW." *International Journal of Man-Machine Studies* 34 (3): 435–452.

Hakken, D. 1994. *Cultural Constructions of Computing: Lessons for the Nordic Computing Project*. Ithaca, NY: SUNY Institute of Information Policy.

Hales, M. 1994. "Where are Designers? Styles of Design Practice, Objects of Design and Views of Users in Computer Supported Cooperative Work." In *Design Issues in CSCW*, eds. D. Rosenberg and C. Hutchison, 151–77. London: Springer Verlag.

Hall, E. T. 1976. *Beyond Culture*, Garden City, N.J.: Doubleday.

Hannerz, U. 1992. *Cultural Complexity*. New York: Columbia University Press.

Heaton, L. 1997. *Culture in Design: The Case of Computer Supported Cooperative Work*. Ph.D. diss., Université de Montréal, Montreal.

Hofstede, G. 1980. *Culture's Consequences: International Differences in Work-Related Values*. Beverly Hills, CA: Sage.

Ishii, H. 1990. "TeamWorkStation: Towards a Seamless Shared Workspace." *Proceedings of CSCW'90*, 13–26. New York: ACM.

Ishii, H., and K. Arita. 1991. "ClearFace: Translucent Multiuser Interface for TeamWorkStation." In *ECSCW '91 Proceedings*, eds. L. Bannon, M. Robinson and K. Schmidt, 163–174. Amsterdam: Kluwer.

Ishii, H., and M. Kobayashi, M. 1992. "ClearBoard: A Seamless Medium for Shared Drawing and Conversation with Eye Contact." *Proceedings CHI'92*, 525–532. New York: ACM.

Ishii, H., M. Kobayashi, and J. Grudin. 1993. "Integration of Interpersonal Space and Shared Workspace: Clearboard Design and Experiments." *ACM Transactions on Information Systems (TOIS)* 11 (4): 349–375.

Ishii, H., and N. Miyake. 1991. "Toward an Open Shared Workspace: Computer Video Fusion Approach of TeamWorkStation." *Communications of the ACM* 34 (12): 37–50.

Ito, S. 1986. "Modifying Imported Technology by Local Engineers: Hypothesis and Case Study of India." *Developing Economies* 24: 334–348.

Jackson, M. H. 1996. "The Meaning of 'Communication Technology': The Technology-Context Scheme." *Communication Yearbook 19*. Thousand Oaks, CA: Sage.

Law, J., and W. Bijker. 1992. "Postscript: Technology, Stability and Social Theory." In *Shaping Technology, Building Society*, eds. W. Bijker and J. Law, 290–308. Cambridge, Mass.: MIT Press.

Mackenzie, D., and J. Wacjman, J. 1985. *The Social Shaping of Technology: How the Refrigerator Got its Hum*. Milton Keynes: Open University.

Madu, C. 1992. *Strategic Planning in Technology Transfer to Less Developed Countries*. New York: Quorum.

Mouer, R., and Y. Sugimoto. 1986. *Images of Japanese Society: A Study in the Social Construction of Reality*. London: KPI Ltd.

Okada, K., F. Maeda, Y. Ichikawa, and Y. Matsushita. 1994. "Multiparty Videoconferencing at Virtual Social Distance: MAJIC Design." *Proceedings of CSCW'94*, 385–394. New York: ACM.

Orlikowski, W. J. 1992. "Learning From Notes: Organizational Issues in Groupware Implementation." *Proceedings of CSCW'92*, 362–369. New York: ACM.

Orlikowski, W. J., and D. C. Gash. 1994. "Technological Frames: Making Sense of Information Technology in Organizations." *ACM Transactions on Information Systems*, 12 (2): 174–207.

Poole, M. S. and G. DeSanctis. 1990. "Understanding the Use of Group Decision Support Systems: The Theory of Adaptive Structuration." In *Organizations and Communication Technology*, eds. C. W. Steinfeld and J. Fulk, 173–193. Newbury Park: Sage.

Raman, K. S., and R. T. Watson. 1994. "National Culture, IS, and Organizational Implications." In *Global Information Systems and Technology: Focus on the Organization and Its Functional Area*, eds. C. P. Deans and K. R. Karwan, 493–513. Harrisburg, PA: Idea Group.

Rogers, E. M., and F. Shoemaker. 1971. *Communication of Innovation: A Cross-Cultural Approach*. New York: Free Press.

Star, S. L., and J. R. Griesemer. 1989. "Institutional Ecology, 'Translations' and Boundary Objects: Amateurs and Professionals in Berkeley's Museum of Vertebrate Zoology," 1907–39. *Social Studies of Science* 19: 387–420.

Stewart, E. C. 1987. "The Japanese Culture of Organizational Communication." In *Organization-Communications: Emerging Perspectives II*, ed. L. Thayer, 136–182. Norwood, NJ: Ablex.

Watson, R. T., and J. C. Brancheau. 1991. "Key Issues in Information Systems Management: An International Perspective." *Information and Management* 20 (3): 213–223.

Wenger, E. 1994. *Communities of Practice*. New York: Cambridge University Press.

Winner, L. 1993. "Upon Opening the Black Box and Finding it Empty: Social Constructivism and the Philosophy of Technology." *Science, Technology, and Human Values* 18 (3): 362–378.

Internet Discourse and the *Habitus* of Korea's New Generation

Sunny Yoon

Introduction

Many people stress the hopeful vision of the Internet as achieving a communication revolution in the contemporary world. They believe that the Internet will break up current structures of inequality in twentieth century mass communication. Proponents portray the Net as a means of expanding participatory democracy, equality, and diversity. By providing an interactive communication medium, they claim, the Internet will allow people to participate in decision-making processes and to produce messages instead of simply consuming the offerings of mass media. Enthusiasts also say that the Internet offers free services that will foster information-sharing with everyone in the world, based on a communitarian spirit.

Recently, however, critical scholars have begun to point out more dangerous possibilities of misusing the Internet. Howard Besser (1995), for example, observes that the Internet can be led in the wrong direction if it is conceived of as an Information Superhighway. Besser argues that this conception, as it stresses commercial uses, may suppress the development of the Internet as supporting democratizing dialogues in an electronic public sphere. According to Besser, commercialism on the Information Superhighway transforms the Internet in four key ways:

1. from flat fee to pay-per use;
2. from an orientation towards the user as producer to the user as consumer;
3. from information to entertainment; and

4. from small niche audience to mass audience (international) (61–67).

These indicators demonstrate that the Internet as an Information Superhighway does not necessarily mediate a revolutionary form of communication. Instead, increasing commercialism may well destroy any new possibilities of constructing the public sphere on and through the Internet.

Following Besser's argument, in practice, the Internet may neither promote democratic communication nor construct a virtual public sphere. It may be that the Internet leads to commercializing information and to alienating a powerless audience. If this is true, it does not mean that the Internet is a useless communication technology, but rather, in the existing social context, the Internet may be used as a controlling mechanism by favoring power and capital, instead of protecting democratic participation and equality.

Power as involved in the Internet is not necessarily a repressive one. Using Foucault's concept of power (1979), power on the Internet can be seen as positive, one that mobilizes people's voluntary participation in the virtual world system. In contrast with more traditional notions of power in the form of a central force that seeks to impose given laws, behaviors, etc., there is no power center that directs and orders people how to use the Internet. Rather, the Internet is scattered all around the world among fifty million users who cannot be tightly controlled. On the Internet, power is exercised in a webbing mode, as Foucault argues. Users are not guided by a linear hierarchy, but they themselves participate in producing authority at the every corner of the world while molding the web of power.

This does not mean that power on the Internet is formless and non-directional as postmodernists assume. To be sure, Bourdieu understands power here to have no source nor any purpose for its use, nor are there any criteria we can invoke to critique power. At the same time, however, power contains a stain, a remnant, of structural force. Bourdieu attempts to examine the structural impact on diverse practices of people in the postmodern world. His concept of *habitus* explains how structural power is constantly reproduced by individuals at the micro-level. Even though some scholars define Bourdieu as a structuralist (Bidou 1988; Hradil 1988), the concept of *habitus* is similar to poststructuralist micro-politics (Foucault 1980).

While Foucault uses words such as power and discourse that are intentionally fuzzy because of his poststructuralist position, Bourdieu's *habitus* implies a clearer, more concrete concept.[1] Bourdieu does not hesitate to call reproduction of the power mechanism cultural capital. In contrast with most poststructuralists or postmodernists who tend to avoid using Marxist terminology such as capital, Bourdieu extends the concept of capital into the areas of culture and symbols. Although postmodernists see a potentially dangerous modernist view in Bourdieu, he is critical of structuralists including Foucault and Derrida as well as Lévi-Strauss and Saussure. The point that Bourdieu makes against structuralism is that it eliminates the will power of subject. By contrast, Bourdieu emphasizes the importance of human practice in the reality of everyday experience.

In order to scrutinize human practice, Bourdieu focuses on the symbolic power of discourse. This is similar to Foucault's conceptual framework of discourse analysis. The uniqueness of Bourdieu, however, is that he underscores strategy, whereas other poststructuralists avoid presenting and analyzing future plans and acting strategies. Bourdieu's *habitus* is a kind of strategy that is accumulated and internalized by individuals' experiences. It may sound deterministic from the poststructuralist point of view, but what Bourdieu emphasizes is the importance of individuals' practices in their everyday lives as these build up society and history. As Bourdieu illustrates, *habitus* has an orchestra effect (1977). Diverse interests and experiences of individuals are integrated into a structural practice, or *habitus*. Yet, this concept of a *habitus* is far from linear or deterministic in a simple sense: though they are in a *habitus*, individuals can adopt distinctive strategies (Bourdieu 1979).

Compared to Foucault's inclusive concept of power, Bourdieu clearly views power as capital. Bourdieu's concept of capital, however, does not stop at Marxist materialism. He takes into consideration cultural capital as well as economic capital. Cultural capital includes symbolic and non-institutionalized power. Cultural capital contributes to maintaining the existing authority by generating *meconnaissance* (misconsciousness) of the majority (Bourdieu 1982). It legitimizes existing authority, however arbitrary such authority may be. Cultural capital induces individual practices in conformity with *habitus*. This conformity is not simply a one-way repression, however: there are constant symbolic struggles in a society.

Education and language are the most prominent examples of cultural capital. Even though education and language are practiced through the voluntary participation of individuals, Bourdieu argues that they contain symbolic violence. This is similar to Foucault's concept of positive power. In the contemporary world, power is not exercised by means of a repressive mechanism. Rather, individuals voluntarily participate in power reproduction while subjugating themselves to the dominant discourse (Foucault 1979). Their bodies and souls are trained to be docile in the process of developing technology and science. The dominant discourse thus becomes the dominant power/knowledge (135–69). Despite the risk of oversimplification, Bourdieu identifies Foucauldian power/knowledge as a *meconnaissance* that exercises symbolic violence in people's lives.

Symbolic violence is not practiced by visible repression or by mutual communication (consciousness). According to Bourdieu and Foucault, education and literacy are not means of enlightenment, but of misconsciousness and control—of *habitus*. This *habitus* includes not only institutionalized education, but all kinds of rules and orders in daily life which socialize the new generation in a certain way; how to walk, sit, talk, eat, etc. Bourdieu (1982) calls it physical *hexis*, which is similar to the docile body in Foucault's *panopticon* (Foucault 1979). These small rules constitute the *habitus* which reproduces existing power. Hierarchical power is maintained by social distinction. According to Bourdieu, class relations are also intertwined with social distinction and symbolic struggles.

This paper will look at the concrete process of reproducing power by means of *habitus* (Bourdieu) and micro-politics (Foucault) in the virtual world of the Internet. I argue that the Internet exercises symbolic or positive power on the new generation by guiding educational rules and linguistic manners. Contrary to the claims that the Internet democratizes, my analysis will show that the Internet is not free of power or symbolic violence in Bourdieu's sense. As Foucault and Bourdieu argue, it is not visible violence or repressive power that is involved in the Internet. Instead, it is a more subtle form of power practice that induces both resistance and subjugation. On this view, it would appear that the Internet can lead either to democratic communication or (cultural) capitalist dominance. Consequently, favoring either direction is pointless unless one looks at the concrete process of how the Internet functions as the *habitus* of people in their everyday lives.

I attempt to uncover this concrete process through research that examines the everyday use of the Internet by Korean youngsters and its cultural meaning for them. First, it is clear that Korean journalism has had a great impact on mobilizing Internet use; in particular, newspaper companies lead the way in encouraging Internet use among school students. So I begin by examining the discourse employed in Korean print media concerning the Internet, partly as a way of uncovering the themes and expectations which enter into the *habitus* concerning Internet use in Korea. Second, I present an ethnographic study of sixteen Korean students in elementary, junior high, and high schools. Most of these students were novice Internet users whom I interviewed for two months in 1996.

Generally, it appears that through symbolic power, the Internet integrates people at the margin of the world into the *habitus* of the virtual world system. The Internet educates Korean younger generation through guiding rules and the pre-existing order of the virtual world system. In Bourdieu's terms, it constantly implants a certain taste and knowledge by means of social distinction and *meconnaisance*. The controlling mechanism of the virtual world system is different from political intervention and physical exploitation in the modern world system (Wallerstein 1979). In the contemporary world, it is not economic, but cultural power that controls people. Cultural capital is the most effective controlling mechanism in the virtual world system. It efficiently affects people—in this case, the younger generation as a minority in Korea—at the very corner of the world. In the virtual world system, the Internet provides a legitimate way of using language and imparting knowledge. People voluntarily habituate themselves to the rules of the Internet while believing it to be a treasure island of the most advanced information and a democratic means of communication. This voluntary participation is not false, but is based on *meconnaissance* or the power/knowledge of science and technology.

Specifically, I begin by measuring and analyzing discourse concerning the Internet as presented by Korean journalism. In Korean society, the Internet is rapidly disseminated partly because of a social mobilization effected by journalism and political propaganda. Competition among the most widely circulated Korean newspapers has led to the social movement of disseminating the Internet: Kidnet on *Chosun Daily*, The Internet Youth Camp (IYC) on *Dong-A Daily*, and The Internet in Education (IIE) on *Joong-Ang Daily*. *Hanguerae*, the most progressive and critically-oriented newspaper, also provides extensive coverage of the Internet. These newspapers have special sections on the Internet and information technology every week. Kidnet of *Chosun Daily* is a movement for distribution of the Internet to elementary school students, whereas IIE of *Joong-Ang Daily* is for junior and high school students and IYC of *Dong-A Daily* is for college students. Competition in the newspaper industry limits each other's territory among the Internet users.

Secondly, this paper adopts an ethnographic approach in order to scrutize Internet use in people's everyday life. It will examine

Internet users in the context of their subculture. Young Koreans, or "Generation X," use the Internet while being affected by dominant symbolic power and, at the same time, exercising resistance to the dominant and imminent power of the older generation. In reality, however, they are the minority group. They are not only at the margin of capitalist power, but also at the margin of symbolic power. Examining the process of encouraging young Koreans to take up the Internet will demonstrate the way symbolic power is exercised in the virtual world system. The Internet offers a course of cultivating the *habitus* of the new generation that relies on symbolic power. It provides legitimate guidelines for how to talk, behave and even think in the virtual world system. By stimulating the distinctive taste of Western culture, the Internet integrates people at the very margin into the virtual world system.

Korean Journalism: Leading the Internet Movement

Since March 1996, the Korean newspaper industry has been led through competition to disseminating information about the Internet and its use to the younger generation. *Joong-Ang Daily* started to organize the IIE (The Internet in Education) movement on the 3rd of March, one day prior to *Chosun Daily*'s starting its Kidnet movement.[2] *Joong-Ang*'s plan was to establish the Internet infrastructure in junior and high schools, whereas *Chosun* attempted to disseminate the Internet among elementary school students. Later, *Dong-A Daily* also organized IYC (The Internet Youth Camp) for college students.

By mobilizing these social movements devoted to Internet use, Korean newspaper companies have promoted social support of the Internet among students. Companies encourage support from private companies and public agencies as well as volunteers. Kidnet of *Chosun Daily* seems to evoke both considerable impact and controversy because children are the target of the movement. Kidnet is connected to the Global Youth Network, which originated in the US. *Chosun* attempts to support introducing computer hardware and network systems in elementary schools by encouraging financial support from private companies. It also organizes volunteers who can teach Internet skills to grade school students. In addition, *Chosun Daily* has designed many ceremonial occasions for Kidnet.

Joong-Ang Daily also has diverse plans for IIE. It has selected sixty middle and high schools, and promotes support from public and private institutions. In addition to providing selected schools with computer hardware, *Joong-Ang* offers to design the Internet home pages for schools free of charge. It also provides diverse instructional programs for IIE.

Dong-A Daily started IYC about two weeks later than *Chosun* and *Joong-Ang*. *Dong-A* offers free Internet classes for universities nationwide. It also has a program for grading the level of informatization of Korean universities. In fact, it evaluates the quality of university education primarily in terms of the level of informatization.

Relying on its power as a tool of mass communication, Korean journalism stands out as a primary force for adopting the Internet in education. It leads this social movement, in part, by offering fantastic dreams of new possibilities to Internet users. As an indication of the extent to which the Internet is represented in Korean journalism, Table 1 shows newspaper reports regarding the Internet. One of the most progressive papers, *Hanguerae*, and one in the conservative side, *Dong-A Daily*, are compared in Table 1.

This table gives a general idea of how Korean journalism portrays the Internet.[3] According to Table 1, there are 132 simple informational reports on *Dong-A*, and 92 on *Hanguerae*. Simple informational reports consist of announcing new services, URLs, Internet instructional programs and so forth. Next to simple information, Internet business is the most frequently reported issue on both

Table 1
Reports of the Internet in Korean Journalism

Title	Dong-A	Hanguerae
Simple informational reports	132	92
New technology	23	10
Foreign industry	39	35
Domestic industry	49	13
Foreign policy	24	10
Domestic policy	26	19
Cultural impact	13	0
Social movement	13	6
Informatization	26	13
Indecency	10	2
Internet use	12	2
Education	25	5
IYC movement	95	—
Others	51	8
TOTAL	538	215

Hanguerae and *Dong-A*. It constitutes 22% of 215 reports on *Hanguerae* and 16% of the total 538 articles on *Dong-A*. (Simple informational reports also contain a great deal of business-oriented information.) *Dong-A* has slightly more reports on domestic business than foreign business, whereas *Hanguerae* has three times more reports on foreign Internet business than domestic business. *Dong-A* tends to introduce more domestic business activities than *Hanguerae*, which has more critiques of Internet business. The latter, for example, criticizes media conglomerates, potential invasion of privacy, and over-competition among the domestic media industry. Overall, Korean journalism addresses Internet business more frequently than any other categories.

Policy-related news regarding the Internet is the third-most frequently reported issue. On *Hanguerae*, 13% of the total consists of policy issues, while policy issues make up only 9% of *Dong-A*'s reporting. These newspapers also address Internet regulation and security problems.

Compared to economic and policy interests, cultural aspects and grassroots citizens' movements are less frequently reported. Regarding cultural and social movement issues, *Dong-A* has 4% of the total, and *Hanguerae* has 2%.[4] This directly contradicts the optimistic theoretical arguments regarding the Internet as a democratizing medium (see Tehranian 1990). According to these arguments, the Internet organizes citizens' or "netizens" ("Net citizens") movements on the global scale by providing a participatory and non-discriminatory mechanism of communication. However, discourse concerning the Internet in Korean newspapers represents the Internet as primarily business-oriented. Further contradicting the theoretical vision of teledemocracy, the Internet is also not politically mobilized: policymakers are interested only in regulation and surveillance issues, instead of developing it as a channel of democratic communication.

As a single agenda, indecency is one of the most frequently cited problems. *Dong-A* has more articles on indecency compared to *Hanguerae*. Although indecency on the Internet is an important issue, the way that journalism represents indecency is problematic. While criticizing indecency, some articles ironically show indecent materials and their URLs (*Dong-A*, April 15, 1996). They stimulate readers' curiosity instead of providing thoughtful criticism. This practice amounts to simple sensationalism in Korean journalism—a sensationalism that is also rooted in prevailing commercial interests.

Table 2 illustrates the Kidnet movement reported by *Chosun Daily*. Although Korean journalism usually emphasizes objectivity, *Chosun*'s Kidnet is unique. *Chosun* attempts to mobilize a social movement by eliminating journalistic objectivity altogether. Many articles regarding Kidnet have no substantial information, but simply highlight positive elements supporting the Kidnet movement, e.g., large numbers of reports concerning emotional reactions to and praise of the Kidnet movement.

According to Table 2, financial and technological support by industry is most frequently reported—almost 20% of the total 175 articles. *Chosun Daily* encourages business to donate computer hardware and software. *Chosun*, in return, advertises industrial supporters on its regular news. *Chosun Daily* also holds many ceremonial occasions as part of the Kidnet movement. Describing the dissemination of the Internet and Kidnet ceremonies are the second most frequently reported articles—18% and 16% of the total, respectively. In order to encourage the Kidnet movement, *Chosun* organizes volunteers and teachers' groups. Articles related to volunteers and teachers' activities are the third largest reported issues. Also, *Chosun Daily* stresses the international connections of the Kidnet movement (9%). *Chosun* has frequently cited the Global Youth Network (GYN) and claims that Kidnet is a part of the GYN movement. According to these reports, Kidnet is a symbol of globalization and international competitiveness.

Table 2
Kidnet Reports of *Chosun Daily*

Title	Number
Program production	2
Simple informational reports	14
Industrial support	36
Citizens' support	20
Teachers' activity	9
International relations	14
General reaction	5
Internet dissemination	31
Public policy	8
Kidnet ceremony	28
Education	8
Others	0
TOTAL	175

While celebrating the Kidnet movement like a ceremonial event, *Chosun Daily* does not sufficiently consider the use value of either the Internet in general or Kidnet in particular. There are very few articles discussing educational usages and problems of young Korean Internet users. Nor does *Chosun* make much effort to help Korean children become Internet producers instead of simply consumers: according to Table 2, there are only two articles regarding the production of educational programs on Kidnet.

In short, Korean journalism does not consider the use value of the Internet, nor does it attempt to utilize the Internet as a participatory communication medium. On the contrary, it contributes to commercializing the Internet by advertising industrial support and promoting competition among the newspaper industry.

Confronting journalistic competition with regard to Internet dissemination, citizens are organizing grassroots social movements in cyberspace. On Chollian, the most widely circulated PC commu-

Table 3
Public Bulletin Boards on Kidnet

Title	We Oppose Kidnet	Kidnet Corner
Opposition		
Media conglomerate	23	10
Advertising	4	1
Negative effect on education	13	10
Cultural dependency	2	3
Demise of kid culture	10	6
Better education first	30	30
Language problem	7	6
Harmful information	3	2
Inequality	11	7
The Internet as useless	10	2
Developing hardware, software	5	23
Other	4	20
Support		
With no reservation	3	15
Educational materials	4	2
Hopeful ideas	6	6
Creative education	8	5
Early education	6	17
Opinions of Reporters	5	5
Others	32	113
TOTAL	186	284

nication network in Korea, there are public discussions regarding Kidnet. Public bulletin boards have been organized. One of them is called "We Oppose Kidnet," and the other is the "Kidnet Corner." Table 3 illustrates citizens' opinions concerning Kidnet.[5]

On the bulletin board "We Oppose Kidnet," the most frequent point of criticism attacks the commercialism of Kidnet. Out of a total of 186 articles, 16% criticize the commercial interests of a media conglomerate such as *Chosun Daily* and its use of Kidnet as an advertising and marketing tool. In particular, some people contend that commercializing the Internet in these ways leads to unequal access to information.

Others point out that Korea should establish a better educational structure before stressing Kidnet (18%). Some more aggressively decry the negative educational effects of Kidnet (7%). Additionally, the potential cultural impacts of the Internet are considered important issues. Some people are concerned that emphasizing Internet use will suppress children's outdoor culture of play and instead constrain them within small rooms (5%). Two articles also bring up the problem of cultural dependency on foreign influence.

Some people worry that indecent information will negatively affect elementary school children. The language barrier to Kidnet users is also often pointed out since the most information found on the Internet is in English.[6] Some people argue that the Internet is in fact worse than useless for primary school students: they claim it makes students lazy and that they lose touch with what is of real value in their lives (5%).

Responding to such critical opinions, some people discuss the positive side of Kidnet on the same bulletin board. They claim Kidnet has beneficial effects on Korean education (6%). Journalists also participate in the discussion. There are five articles by Kidnet reporters on the bulletin board.

Table 3 also makes clear that at "Kidnet Corner" there are more diverse opinions than on "We Oppose Kidnet." More supportive opinions are presented, and people are more interested in technological issues rather than social ones. Compared to "We Oppose Kidnet," dissenting opinions on "Kidnet Corner" do not deny the value of the Internet in education: they only question the methods of the Kidnet movement.[7]

Two public bulletins argue that Korean schools should first improve educational conditions before utilizing the Internet (thirty articles each). Without establishing the appropriate infrastructure, they argue, organizing Internet use will be less useful for students.

Facing the Kidnet movement mobilized by Korean journalism, people organize public discussions in order to resist its commercialism and what they fear will be the negative cultural impacts of the Internet on young Koreans. Although it does not represent the opinion of the all Koreans, it is valuable to look at these opposition readings.

The point of this research is not to attack Korean journalism, but to examine its impact on everyday life of the Internet users in Korea. Internet users are mobilized and affected by newspaper reports of the Internet. Journalistic discourse about the Internet contributes to the *habitus* of the Internet users. Users voluntarily participate in conforming to the dominant discourse, or *habitus*, without "being the product of obedience to rules . . . presupposing a conscious aiming at ends" (Bourdieu 1977, 72). They are not aware of being affected by power/knowledge as represented by the Internet. They use the Internet based on a kind of mythical belief, or *meconnaissance*. In particular, these users conform to the discourse of scientism and technocratism—the beliefs that science will discover objective truth in evolutionary stages of human history, and technology will set people free from physical and mental constraints; users make themselves believe that the Internet will bring about a futuristic dreamworld. They also believe that Internet users will be leaders of the "Information Society" of the twenty-first century.

But it is precisely these scientistic and technocratic myths of inevitable advance that Bourdieu characterizes as *meconnaisance*. We should note, however, that *meconnaisance* is different from false consciousness or ideology as Marx understands these. *Meconnaisance* is not externally imposed by repressive authority. Rather, it is internalized in peoples' minds and it produces reality by means of human practice. In these ways, it constitutes a *habitus* that constructs history and society. In order to look at the *habitus* of Internet use in Korea, I now turn to an ethnographic study of young Korean Internet users.

The *Habitus* of Internet Users and Subculture of Korea's New Generation

This section examines the *habitus* of young Koreans who use the Internet in their everyday lives. The new generation of Koreans is ambivalent regarding this new technology: using the Internet may mean both conformity to the dominant values of the society and at the same time resistance against the authority of the older generation.

On the one hand, the Internet is taken to be the sign of the most advanced information technology, a technology that will lead to prosperity and convenience in the society of the future. Korean youngsters follow this maintream idea—one that is consistently represented in Korean journalism—as Table 1 shows.

On the other hand, for Korean young people the Internet is also a symbol of youth and resistance. They experience new authority by using the Internet. This research will uncover these diverse meanings of the Internet for Korean young people. It portrays the *habitus* that contains both structural conformity and diverse forms of resistance.

I have conducted an ethnographic study examining the *habitus* of the Internet users. I interviewed sixteen students in elementary, middle, and high schools in 1996 (see table 4).

I conducted participatory observation of these students in their homes and schools. I had individual interviews with each of them, and group discussions with the students according to level—i.e., grade, middle and high school students. Both individual and group interviews were tape-recorded and then transcribed immediately.

Table 4
List of Interviewees

School	Name	Age	Gender	Computer Use (hours)*
High	C	16	F	3 (0)
	J	16	F	1/2 (4)
	H	16	M	1/2 (3)
	O	16	M	1/2 (3)
	Y	16	M	1 (3)
	L	16	M	1/2 (2)
	A	17	M	1/3 (2)
Junior High	G	15	M	1 (4)
	E	15	M	2 (8)
	S	15	M	1 (3)
	K	15	M	1 (3)
Elementary	B	12	M	1/3 (1)
	C	11	M	1/3 (1.5)
	X	11	M	1/2 (1.5)
	M	12	F	1/2 (1)
	N	11	F	1/2 (2)

* Average computer and the Internet use hours during academic year.[8]

This approach thus attempts to present possible meanings of the Internet for Korea's new generation through their own interpretations and voices. This research makes no attempt to represent all Korean students, nor to generalize its findings. It is rather a special case, one limited to these sixteen students. It is meaningful, however, to examine in this way the concrete process of reproducing *habitus* in individuals' everyday lives. Using the Internet is not simply a mechanical process of extracting useful information: rather, this use contains diverse meanings for the young people studied.

Korean youngsters use the Internet not because they need information, but because they want to be seen as advanced. Most of the students I interviewed have a positive attitude towards the Internet.[9] Yet most of the interviewees confess that they do not need the information found on the Internet. For example, "E" asked me in response to one of my questions, "What would be the usefulness of the Internet? I am not doing business of any kind." "A" and "B" expressed a similar idea in interviews. Still, they fantasize about using the Internet because they want to be seen as elites and new explorers of advanced technology. For example, "M" and "N" (elementary school students) said that they would like to use the Internet because they wanted to be viewed differently. "C," who does not use the Internet, also said that she had spent a lot of time using PC communication because it made her look more advanced. In turn, she perceived other students who did not use PC communication as "primitive." Now, she spends less time with PC communication because "so many people use PC communication, there is no scarcity value. I do not want to spend so much time with them." ("C's" case may apply to more Internet users in the near future: up to this point, since there are only a small number of the Internet users, they are proud of being the front-runners of Internet use.)

Although Korean journalism emphasizes the importance of the educational use of the Internet by mobilizing social movements such as Kidnet, IIE and IYC, the use value of the Internet is in fact rarely questioned. By contrast, the students I interviewed acknowledge that the Internet does not have much use value with regard to their specific needs. Still, most of them are satisfied with the Internet because it can give a positive image of their identities.

I would argue, however, that this is not false consciousness. In fact, Korean students experience new identities in front of computers. Although they are a minority and are controlled by adults, in cyberspace the young generation defeats the old one. Even elementary school students, "B" and "C," teach their fathers how to use computers and guide other adults in cyberspace. "E" in middle school, for example, designs all kinds of official documents at school at teachers' requests. He is much better than his teachers in the area of computer use. Similarly, "E," "G," "K," and "S" are proud of building their schools' Web homepages, thereby expressing their own creativity. No higher authority orders them to do something, because the students are the ones who know best regarding the Internet. Authority figures, including the old school principal, comply with these students' ideas because the older generation is ignorant compared to these kids.

In short, the Korean new generation experiences an alternative identity in cyberspace that they have never achieved in real life. The hierarchical system of ordinary social reality turns up side down as soon as Korean students enter cyberspace. In interviews, most students claimed that the Internet opened a new world and new excitement. This is not only because the Internet has exciting information, but also because it provides them with a new experience and an alternative hierarchy. It is something of an experience of deconstructing power in reality, especially in Korean society, which is strongly hierarchical and repressive for young students.

It is the *habitus* of Korea's new generation that makes them partly comply with the dominant discourse and partly resist the old authority system. Using the Internet is a strategy of the new generation; a *habitus* affects people's minds and living patterns through its symbolic power, and at the same time it allows for diverse strategies of resistance. Relying on the cultural power of new technology, young Korean students attempt to break up the hierarchy of old authority and experience their new identities in the cyberspace.

But constructing an alternative identity for the new Korean generation cannot be accomplished without struggle. As Bourdieu points out, instead of imposing a linear order of structural force, *habitus* pertains to struggles on the symbolic level. First, Korean students face the opposition of their parents and teachers. Although Korean journalism emphasizes the futuristic role of the Internet, the old generation is suspicious of the new machine, the computer. In Korean society, the only value that students must fulfill is study. Interviewees complained that their parents and teachers demanded that they limit their "play time" on the computer, at least to a specified amount. Adults usually argue that students can use the computer as much as they want in the future when they enter universities. Confronting the antagonism of adults, young Koreans resist the value system of the older generation. In interviews, most students argued that their use of computer and the Internet was not sufficiently supported at their homes and schools. "K" claimed that the older generation did not understand the "computer generation"—including him—due to their lack of experience with the new technology. "H" said that his parents complained about his "obsession" with computer games whenever he used computers. But according to "H," since most contemporary computer program menus consist of graphics, his parents believed that he was playing computer games all the time. These students laugh at the ignorance of the older generation, while also confirming their superior authority over the older generation with regards to computers. They are not inclined to communicate with the older generation, or to persuade them to adopt new views or values. Most of them agree with the statement, "It's useless to talk to them. I just go my own way."

Second, a more serious problem that young Koreans have on the level of symbolic struggle is the language problem. In my interviews, all students noted that the most difficult problem with using the Internet is understanding English. On the Internet, language constitutes the *habitus* as the most efficient power mechanism in the virtual world system. Without questioning the power of language on the Internet, Korean youngsters take it for granted that they have to adjust to the linguistic "grammar" of the Internet. Reading Internet hypertexts in English consumes a great deal of their money and time. Most students pointed out that they had to use a dictionary to decode messages, which costs a lot of time and Internet connection fees. Most of the time, they skimmed through the text because they had difficulties in reading it.

However, young Koreans do not consider the linguistic practice of the Internet as an imperialist power. Instead, these students try to comply with the linguistic rule of the Internet. In cyberspace, Korean students find themselves powerless and ignorant. "L" said that his identity in the cyberworld is "like a tiny grain of sand in a big ocean," and "A" considered himself to be a "free-rider." "B" (in elementary school) said that he was "only a powerless consumer," "C" defined himself as "a parasite who makes no contribution." Although these students are disturbed by the language problem, they find fault with themselves because they cannot fluently adapt to the *habitus* of the virtual world system. Young Koreans are hopeful, however, that they will contribute to the Internet corpus in the future when their English is improved.

In the virtual world system, language exercises symbolic power. It guides the new generation of Koreans regarding how they talk, write, behave and think. It is a symbolic violence that constrains Korean youngsters in their reading, writing and knowing information. As symbolic power, language constructs a *habitus* that makes people accept what they believe to be the legitimate way of behaving and thinking in the virtual world system. It is violent in Bourdieu's sense, but it symbolically influences people without being noticed.

In the cyberworld, English is cultural capital. In a group discussion, "E" and "S" contended that leaders on the Internet are those who can speak English well, and that they "are dominating the Internet, even though it is not totally controlled." In the virtual world, language marks distinctions between classes, cultures, generations and mastery of diverse kinds of knowledge. Monetary capital marks a class distinction between the bourgeois and the proletariat, both domestically and internationally; in the modern world system, language as cultural capital exercises symbolic power over the cultural have-nots in the virtual world system. English greatly hinders the access of Korean students to information on the Internet. Cultural capital provides the most efficient mechanism to induce the voluntary subjugation of people in the world at the micro-level, and it also contributes to maintaining and accumulating economic capital in the long run. It constitutes the *habitus* that domesticates people's lives at the margin and legitimizes *meconnaisance*.

Third, the new generation of Koreans is involved in symbolic struggle in their resisting the negative side of the Internet. Although young Koreans adopt Internet use as a strategy to practice their new identity in the virtual world, these students also recognize negative effects of the Internet. Most students pointed out that commercialism on the Internet is a problem. "A" emphasized that business interests were dominating and using the Internet as a marketing tool. "A" and "O" agreed that industries released "free" information on the Internet only because it could no longer be marketed. "C" and "X" (elementary students) said that they would mostly use the Internet as a marketing tool in the future.

In discussion, all interviewees agreed that indecent material on the Internet would bring about the most negative effect on students. "M" and "N" (girls) particularly criticized indecency on the Internet and the boys who looked at it. Most boys, including grade school students, have visited sex sites on the Internet. (In individual interviews, some students denied having visited such sites, but in group discussion, students accused and teased each other regarding their experiences of viewing indecent materials.) All students perceived indecency on the Internet negatively and claimed it had to be stopped.

However, according to these young students, indecency is not the most significant issue on the Internet. Contrary to the exaggerated claims in Korean media, "E" and "S" claimed that students did not spend all their time looking for indecent materials on the Internet. At first, students were curious about these materials because they had heard a lot about them. But "G" contended that while those who did not know much about computers showed interest in indecency, most Internet users usually looked for "other more interesting information." Most students who had looked at indecent material on the Internet pointed out that the Internet was less problematic as compared to other media, such as magazines and videotapes. They resist notions of prohibition by the older generation, even though they agree that indecency on the Internet has negative effects.

Concluding Remarks

Although many theorists believe that the Internet constitutes a revolutionary communication tool, this analysis makes clear that the Internet does not necessarily bring about democratic communication and the development of a virtual public sphere in the contemporary social structure. Rather, power interferes with the potentially democratic uses of the Internet—even though this power, in the form of symbolic power and the *habitus* of Internet users, does not visibly repress Internet users. Adopting Foucault and Bourdieu's conceptual frameworks, I have attempted to uncover this subtle form of power involved in the virtual world system through a discourse analysis of Korean journalism and an ethnographic study of Korea's new generation.

Through symbolic power, the Internet integrates people even at the margin of the world into the *habitus* of the virtual world system. The Internet use of Korea's new generation illustrates that cultural capital such as language and education contributes to reproducing existing power. This power affects both authoritative communicators in the mass society and people at the margins, in their everyday life.

Influenced by the dominant discourse, Korean journalism has initiated efforts to disseminate the Internet in the schools. However, these efforts are frequently criticized by grassroots virtual organizations. They argue against commercialism and negative educational effects of the Internet movements. While leading social movements focused on Internet use, Korean journalism does not sufficiently consider the use value of the Internet, nor does it attempt to utilize the Internet as a participatory communication medium.

The young generation of Koreans is mobilized by journalism. For this new generation, the Internet is both a medium of resistance and a potential tool of upward mobility in the future. They find new identities in the cyberspace. Although they are a minority, one controlled by adults, in cyberspace, the young generation defeats the old one. It is an experience of deconstructing authority in reality.

Constructing an alternative identity, however, cannot be accomplished without struggle. Korean students participate in symbolic struggles while using the Internet as a strategy. The *habitus* of the virtual world system offers diverse possibilities the new generation of Koreans. Young people both conform to the dominant discourse, and at the same time resist the dominant power at the micro-level. This ethnographic study demonstrates the concrete process of symbolic struggles of Korea's new generation.

Notes

1. Bourdieu defines the concept of *habitus* as follows:

> The structures constitutive of a particular type of environment (e.g., the material conditions of existence characteristic of class condition) produce Habitus, systems of durable, transposable dispositions, structured structures predisposed to function as structuring structures, that is, as principles of the generation and structuring of practices and representations which can be objectively "regulated" and "regular" without in any way being the product of obedience to rules, objectively adopted to their goals without presupposing a conscious aiming at ends or an express mastery of the operations necessary to attain them and, being all this, collectively orchestrated without being the product of the orchestrating action of a conductor. (1977, 72)

2. *Chosun* claimed that *Joong-Ang* stole Chosun's plan, and declared IIE first. This was discussed on the bulletin boards.

3. *Chosun* and *Joong-Ang Daily* have more reports on the Internet than *Hanguerae* and *Dong-A*. I will look at more specific issues on the former, i.e., Kidnet on *Chosun* and IIE on *Joong-Ang*.

4. I classified articles regarding teledemocracy and grassroots democracy under the category of "social movement."

5. The online bulletin board, "We Oppose Kidnet" operated from April 1, 1996, until July 19, 1996. In this bulletin board system, one person can represent his/her opinions multiple times.

6. Indecency and language problems are mentioned more than three and seven times, respectively. Since I counted the number according to the major point in each article, some articles which mention many things at once are classified according to the major point.

7. At the bulletin board of "Kidnet Corner," there are more arguments between supporters and opponents than the other one. In the course of their arguments, many people stray from the point: I classified these into "Others."

8. In the junior high school, I was unable to interview any female students. Two female students came, but they did not want to be interviewed.

9. In particular, four middle school students spent the whole day every day making the Internet homepage of their school during this summer break.

References

Besser, H. 1995. "From the Internet to Information Superhighway." In *Resisting the Virtual Life*, eds. J. Boork and I. Boal, 59–79. San Francisco: City Lights.

Boudon, R. 1984. *La place du desordre*. Trans. M. Moon. Seoul: Kyobo.

Bourdieu, P. 1977. *Outline of a Theory of Practice*. Trans. R. Nice. Cambridge: Cambridge University Press.

———. 1979. *La distinction critique social du judgement*. Paris: Editions du Minuit.

———. 1982. *Ce que parler veut dire*. Trans. J. Jun. Seoul: Saemulgul.

———. 1984. *Distinction: A Social Critique of the Judgement of Taste*. Trans. R. Nice. Cambridge, MA: Harvard University Press.

Foucault, M. 1979. *Discipline and Punish: Birth of the Prison*. Trans. A. Sheridan. New York: Vintage Books.

Foucault, M. 1980. *Power/Knowledge*. Trans. C. Gordon. New York: Pantheon Books.

Hradil, S. 1988. *Class and Cultural Practice*. Trans. M. Moon. Seoul: Kyobo.

Lee, S. 1994. "The New Social Theory of Bourdieu." *Society and Mass Communication* 5: 3 (Fall), 79–115.

Tehranian, M. 1990. *Technologies of Power*. Norwood: Ablex.

Wallerstein, I. 1979. *Capitalist World System*. New York: Routledge.

“Culture,” Computer Literacy, and the Media in Creating Public Attitudes toward CMC in Japan and Korea

Robert J. Fouser

Introduction

Two views of computer-mediated communication (CMC) have prevailed since it first attracted public attention in the early nineties. To optimists, CMC frees people from the physical constraints of time and space and the social constraints of race, gender, and class. The computer screen allows people to presented a liberated version of the self to a virtual community of other liberated selves. The optimists (e.g., Rheingold 1993; Connery 1997) extend this view to argue that CMC creates new virtual communities that help expand democracy and level the playing field for participants in the global economy. The pessimists (e.g., Stoll 1996; Turkle 1996), on the other hand, view CMC as the final stage in the dehumanization of society. They argue that anonymity of CMC encourages aggressive communicative behavior—flaming, vulgar language, hacking, and Web pornography—to a greater degree than face-to-face communication. Popularized by journalistic writers such as Rheingold, Stoll, and Turkle, these views have become part of contemporary cultural folklore as the media appeals to images of a “cybertopia” or the fear of a “cyberhell” to stimulate the public imagination.

As Internet use and CMC have spread around the world, these views have a number of “mirror sites” around the world. Two technologically advanced Asian societies, Japan and Korea, offer an interesting look at the reception of CMC as a social phenomenon in industrialized non-Western societies where electronically mediated communication interacts with traditional communication patterns. Both have a large number of computer users and subscribers to commercial on-line Internet providers. Computer and Internet use in both societies has grown rapidly since the early nineties as word processing programs and UNIX-based character sets have improved. The rapid growth in computer and Internet use has attracted considerable media attention in both countries. In the process, CMC has become a firmly established part of popular culture and cultural folklore in both countries.

In this paper, I will compare and contrast media and artistic views of CMC as a new pop culture phenomenon in Japan and Korea. From this discussion, I will elucidate important trends that help define the media perceptions of CMC that have emerged in these two countries. Finally, I will conclude with a discussion of which theory—a cultural theory or a computer literacy theory—better accounts for observed differences between Japan and Korea in the early reception of CMC; the better theory, in turn, will help us more fully understand what factors will influence subsequent perceptions of CMC in both countries.[1]

Differences between Japan and Korea

The major national daily newspapers, all of which have extensive Web sites, reveal a number of interesting differences between the two nations. The most obvious difference between, say the *Yomiuri Shimbun*'s Web site (<http://www.yomiuri.co.jp>) and the *Chosŏn Ilbo*'s (<http://www.chosun.co.kr>), the two largest dailies in each country, is the number and type of banners. The *Chosŏn Ilbo* has many large colorful banners, many of which move or change shape. The layout is easier to follow and includes several color photographs. The headlines are large and designed to attract attention much as they are in the print edition. The *Yomiuri Shimbun*, on the other hand, has fewer banners and a more conservative layout. There are fewer photographs, and headlines are smaller. The *Chosŏn Ilbo* also has a site that lets users search back issues of the paper for articles. The entire print edition is presented on the Web site. For the *Yomiuri Shimbun*, users have to pay one thousand yen to register for a user ID and seven hundred yen per month to search back issues. The Web edition does not carry all of the articles that are in the print edition. A survey of other major newspapers in both nations reveals similar findings. Korea has followed the model of many newspapers in the English-speaking world, such as the *Washington Post* and the *Los Angeles Times*, that put the complete paper on the Web free of charge. One difference, however, is that many newspapers in the English-speaking world offering otherwise free access nonetheless charge a fee to download archived articles.

What does this mean for CMC? First, the greater number of banners on Korean Web pages indicates that Korean advertisers believe that the Web is an effective way to reach customers. Although Japan has a large number of Internet users, the paucity of banners indicates that Japanese companies prefer to spend their money on other forms of advertising. Web newspapers thus have a certain mass appeal in Korea that they do not have in Japan. Second, the layout of Korean Web pages is more "accessible" to people who are familiar with the print edition. Japanese Web pages look more "computer-like," which could make them intimidating to readers accustomed to print media or colorful Web sites.

Other frequently accessed Web sites, such as search engines and free e-mail services, reveal interesting differences between Japan and Korea. Both nations have native-language Yahoo! search engines and several other local native-language search engines. The categories in both versions of Yahoo! are the same. Yahoo! Japan offers a wider range of news and other information than Yahoo! Korea. Most of the Reuters news in English appears in Japanese along with Japanese-only news. The financial section of Yahoo! Japan allows for real-time quotes of individual stocks and a variety of financial information. Yahoo! Korea does not have extensive news or financial information because Yahoo! is relatively new to the Korean market. Both nations have home-grown search engines—goo (<http://www.goo.ne.jp>) in Japan and naver.com (<http://www.naver.com>) in Korea—and a number of free Web-based e-mail services modeled after hotmail. The Korean hanmail.net (<http://www.hanmail.net>) boasts over a million users of its free e-mail service.

Population and economics have a direct relationship on the amount of information available on the Web. The Japanese market of 125 million people is larger and richer than the Korean market of 45 million people. According to the *Courrier International* (1998), Japanese is the third most common language on the Web after English and Spanish. Korean is the fifteenth most common language, but it is the only Asian language besides Japanese and Chinese among the fifteen languages listed. Korea has a smaller GNP per capita than Japan: $9,511 versus $33,800. The standard of living for middle-class city dwellers is closer, however, than these figures indicate because the high cost of living and crowded housing conditions in Japan reduce the standard of living. When calculated for purchasing power parity, the gap narrows to $13,990 for Korea and $23,840 for Japan (GNP figures for 1997 from *Asiaweek* 1998; figures for 1998 will show a larger gap between the two nations because of the severity of the economic crisis in Korea). The effects of the high cost of living in Japan on the Internet will be discussed in more detail later.

From this quick look at Web pages and Web browsers in Japan and Korea, two contradictory trends become clear: Japan has more Web pages and Internet users, but the Web has yet to attract the attention of advertisers and the general public. Korea, on the other hand, has fewer Web pages and Internet users, but is rapidly attracting attention of advertisers and the general public, despite a lower level of economic development. A caveat needs to be added here because advertising revenue from on-line newspapers in Korea remains a small percentage of total newspaper advertising revenue. The on-line *Chosŏn Ilbo*, for example, receives 150,000,000 won (about $125,000) in advertising revenue each month (Hong, personal communication). Though this has been increasing each month, it lags far behind the print edition in which one-day full-page advertising spread costs about 50,000,000 won (about $40,000). I was unable to get figures for the *Yomiuri Shimbun* or other Japanese on-line newspapers, but the paucity of ads indicates that the on-line version is designed to advertise the print edition or make money from on-line news services and subscriptions such as the one offered by the *Yomiuri Shimbun*. So far, there has been little open discussion on charging customers in Korea, so it may be that newspaper companies in Korea are hoping that advertising revenue will help offset the cost of maintaining the on-line edition.

Reports on the Web and CMC in the media in each nation reflect the different levels of public interest. A comparison of *AERA* in Japan and *News+* in Korea, two major weekly newsmagazines equivalent to *Time* or *Newsweek*, highlights this difference. *AERA* covers the Internet and CMC periodically and generally from a critical stance. *News+*, on the other hand, has a weekly section devoted to the Internet and often runs short stories on how news events are discussed in domestic chat rooms operated by commercial Internet providers. The tone of most of these articles is positive, particularly regarding the development of Korean-language software and Web sites. In the Korean media, surveying chat rooms has become a way to gauge public opinion and trends in language and popular culture.

Print editions of major newspapers offer an interesting contrast. Korean daily newspapers, such as the *Chosŏn Ilbo*, have a weekly section dedicated to computers and the Internet. Beginning in late 1997, major dailies in Korea, such as the *Chosŏn Ilbo* and the *Chungang Ilbo*, began including the reporter's e-mail address in parenthesis at the end of the article, which gives readers the opportunity to provide direct and immediate feedback on the article. No Japanese newspaper gives e-mail addresses for individual reporters. Articles on the Internet and CMC in major Japanese dailies, such as the *Yomiuri Shimbun* or the *Asahi Shimbun*, appear in various sections of the paper, with technical and economic developments appearing in the business section and social developments in the society and culture sections. The *Nihon Keizai Shimbun*, the Japanese equivalent of the *Wall Street Journal*, has a weekly section on computers and the Internet, but most of the information is from a business and marketing perspective. Advertising in magazines and newspapers also reflects a difference in interest. Japanese magazines and newspapers carry fewer advertisements for computers and commercial on-line services than Korean publications. This difference also extends to posters in subways and other public places.

The greater amount of media attention and advertising in Korea indicates a higher level of public interest in computers, the Internet, and CMC than in Japan. In particular, the Korean media's use of chat-room discussions as a gauge of public opinion and social trends contrasts sharply with the infrequent reports of chat rooms as a social phenomenon in the Japanese media. This, along with the willingness of advertisers to pay for banners on Web pages, indicates the Web and CMC have attracted a higher level of public interest in Korea than in Japan. This is despite the greater amount of information available on the Web in Japanese and the higher level of economic development in Japan. Indeed, Watanabe (1997) correlated the number of Internet host computers per million dollars of GNP and found that Japan lagged behind most Asia countries with only 0.07 computers, whereas Korea had 0.09, Taiwan 0.11, Hong Kong 0.17, and Singapore 0.41. Japan was tied with Malaysia (0.07) for last place on this scale.

Similarities between Japan and Korea

Despite the differences, there are a number of similarities in perceptions of the Internet and CMC in Japan and Korea. One of the most obvious similarities is the commercialization of the Internet in both countries. Unlike the United States, where the Internet had an established place in academic discourse before going commercial, commercial on-line services in Japan and Korea helped spread the Internet in society before it spread to businesses and academic institutions. From the early stages of Internet diffusion, users have been willing to pay for these on-line services in both countries, which has affected public perceptions of the Internet as a tool for business or pleasure, rather than as a tool of information exchange as it was perceived in the United States in the early years of diffusion.

Another similarity between Japan and Korea is the perception of the Internet and CMC in the arts. To artists in both countries, the Internet is a source of entertainment that lets them have fun with their work as they reach out to the public. Visual artists in both countries have embraced the Web through virtual galleries. This action allows them to reach a wider audience because the "gallery" or the "museum" intimidates many people in both countries. The Neo-Pop Japanese artist Majima, for example, used the Internet to sell his works that were displayed as a "convenience store" in a Tokyo gallery (Fouser 1998). This "virtual gallery," which received considerable media attention, was on the Web during the exhibition. Another Japanese artist developed an "art e-mail" software program entitled PostPet that delivers e-mail in an entertaining way. The goal of PostPet is clearly to turn e-mail delivery into a virtual performance on the screen (*Bijutsu Techō* 1997).

Two films, *Haru* in Japan and *Chŏpsok* in Korea, point to similar perceptions of CMC in the arts. Directed by Morita Yoshimitsu, *Haru* is a 1996 film about a chat-room romance. As the chat-room romance develops, the two try to meet in person, but miss each other. Instead, they arrange to glimpse each other from a distance, he in a train, and she in a car by the tracks. *Chŏpsok*, or "The Contact" in English, is a 1997 film about a chat-room romance. As in *Haru*, the two lovers try to arrange a meeting, but end up stalking each other in a coffee shop in downtown Seoul. The failure of the couples in these films to meet implies a fear of direct contact, or perhaps a preference for the anonymity and spontaneity of the chat room. One important difference between the two films, however, is their public impact. *Haru* attracted moderate attention in Japan, but was not a major success. In contrast, *Chŏpsok* was one of the most popular Korean films of all time. Over a million people saw the film on its first run (*Chosŏn Ilbo* 1997).

Interest in the Internet and CMC may not entirely explain the popularity of *Chŏpsok* because at heart the film is a melodramatic love story that was the first of a series of such films in Korea. A better example of the influence of CMC on the arts is the *Ttangji Ilbo*, an on-line parody of the on-line *Chosŏn Ilbo* developed by Kim Ŏjun (*Ttanji Ilbo* 1998). The word *ttanji* in the title refers to a move in traditional Korean wrestling (*ssirŭm*) in which the opponent is blocked and twisted to move in the opposite direction. By turning events on their head, the *Ttanji Ilbo* takes a sarcastic look at events through wry humor and photo manipulation. By modeling itself after on-line *Chosŏn Ilbo*, the *Ttanji Ilbo* also takes a critical look at how established media package information for on-line consumption. By presenting a humorous, but obviously "twisted" version of the news, the *Ttanji Ilbo* forces readers to question the accuracy of on-line editions of established newspapers. In addition to the Web site, contents from the *Ttanji Ilbo* have been published in a series of books. The *Ttanji Ilbo* has received considerable media attention and the Web site and books are popular with college educated persons in their twenties and thirties. To understand the humor of the *Ttanji Ilbo* to the full, however, readers need to be familiar with the on-line *Chosŏn Ilbo* and other on-line papers; the popularity of the *Ttanji Ilbo* Web site and books suggests that they are.

First Impressions, Questions

From the above survey of difference and similarities in media and art perceptions of the Internet and CMC, Korea is clearly more enthusiastic about the Internet as a tool for communication and information exchange. The higher level of public interest and enthusiasm in Korea indicates a more positive stance toward CMC than in Japan. Despite Japan's reputation as a technological powerhouse, the Japanese media and the public harbor doubts about CMC, though, as is evident from the number of introductory TV programs on the Web, they have shown interest in the Web as a deliverer of information. The Japanese public has been quick to embrace other forms of electronically mediated communication, such as the fax and the mobile phone, but not CMC. The important question from this comparison between Japan and Korea is why large segments of the Japanese media and public lack enthusiasm for or are openly critical of CMC. Or, to reverse the question, why are Koreans so enthusiastic about CMC, despite having more limited financial resources?

Explaining the Differences: Two Theories

To answer the above questions, I will posit two theories and evaluate them against evidence from writings about CMC in Japan and Korea. The first theory I will call the "culture theory." This theory holds that distinct cultural differences between Japan and Korea lead to differences in the reception of CMC. This theory is the "made-for-CMC" equivalent of the Sapir-Whorf hypothesis that places culture at the center of the debate. The second theory I will call the "computer-literacy theory." To this theory, the difference between the two nations is a practical issue of differences in computer literary, including typing ability, that affect the image of computers, and hence CMC, in society. This theory takes into account a variety of other practical considerations, such as the cost of computers and on-line time, space in the home for computers, distribution of computers in offices, and the diffusion of competing communication technologies.

In evaluating these theories for Japan, I draw heavily on a four-nation (Japan, Korea, the United States, and Singapore) survey of attitudes toward electronic communication, entitled "Comparative International Study on Electronic Information and Communication," conducted by the Nomura Research Institute in September to November of 1997 (Nomura Research Institute 1998). The survey was conducted in 100 different places in each of the four countries. Researchers interviewed a random sample of 500 adults in Korea, the United States, and Singapore and 1,409 adults in Japan. Subjects were equally distributed between men and women and were limited to the ages of fifteen to fifty-nine. The data for Japan were collected in September 1997 and for Korea, the United States, and Singapore, in a two-month period from October to December 1997. The relatively large sample size and the imposition of similar controls across the four countries give the results considerable validity.

For the culture theory to be valid, there would have to be something in Japanese culture that put it at odds with the Internet and CMC. Much Japanese writing on CMC focuses on problems such as hacking, abusive language, and social alienation. A September 1997 article from *AERA* focused on the "rudeness" of e-mail (Itami 1997). The article called attention to flaming on mailing lists and casual language in messages. It quoted several persons who argued that the ease of sending mail and anonymity of the medium encouraged people to use casual and blunt language. The theme of anonymity appears again in a recent article in the *Yomiuri Shimbun* (Takenaka 1998). The article quotes Professor Okuno of Kansei Gakuin University as saying that media that bring families together in one room, such as the TV, are giving way to individual media, such as the mobile phone, e-mail, and pagers. The article goes on to discuss the development of "individual media" as a danger to the family, and concludes with Professor Okuno's warning that the living room—the center of Japanese family life—may become empty in the near future. Articles such as these suggest that Japanese people fear that CMC will disrupt traditional patterns of behavior that reinforce in-group solidarity through direct human interaction. The importance of group solidarity in the company and schools remains strong in Japanese life. The adoption of individual patterns of communication thus presents a challenge to the solidarity of the group that Japanese institutions may not accept easily. A Japanese computer entrepreneur in Silicon Valley noted in a recent article in the *New York Times* that "Japan is a place where the ties between people are very strong and people like to do business with the people they know personally. The Internet is about networking strangers to talk to each other" (Kotkin 1999).

Another prevailing theme in Japanese writing on CMC is social instability. A recent article in the *Asahi Shimbun*, a major national daily, presented several case studies of women, mostly homemakers, who used e-mail and chat rooms to develop new social networks, including relations with men other than their husbands (*Asahi Shimbun* 1996). The article insinuated that CMC tempts women into cheating on their husbands, an ironic charge in a society where pornographic abuse of women is so prevalent and accessible. The Japanese legal system has recently come to the defense of those who feel victimized by CMC language. In a landmark court case in 1997, the Tokyo District Court ruled that a posting on the "Feminism" BBS in NIFTY-Serve, the largest commercial Internet provider, constituted libel (*Asahi Shimbun* 1997). The case concerned a comment by a male participant in the BBS who said that a frequent female participant who asserted feminist views would "probably end up getting a divorce because she is so stubborn." The woman sued NIFTY-Serve, the manager of the BBS, and the poster of the message. The judge ruled that because a BBS is open to all members of NIFTY-Serve, that the comment "lowered the public image of the victim," which constituted libel. The court said that the BBS manager and

NIFTY-Serve were negligent because they did not delete the message before it was posted and ordered each institution to pay damages of 500,000 yen. In addition to bringing up serious issues of free speech on commercial on-line services, the court's decision reflects the prevailing view in Japan the direct, critical comments in CMC cause harm to the addressee.

Results from the Nomura survey, which are presented in Table 1, reveal a similar apprehension toward and, in comparison with the three other nations surveyed, ambivalence about the Internet. According to the survey, 56.4% of Japanese respondents said that computers and other information technologies would not increase human communication. This figure contrasts with 25% in the United States and 23.6% in Korea. In response to the question "Do you worry about not being able to use a spreading new technology that spreads in society?" 59.6% of Japanese said "yes," whereas as 70.8% of Americans and 72.2% of Koreans said "yes." In response to the question "Do you want to let the world know of your existence and ideas?" 66.8% of Japanese said "no," whereas only 34.6% of Americans and 41% of Koreans said "no." In response to the question about homepages, 71.2% of Japanese said that they did not want to set up a personal homepage. Seventy percent of Americans did not want to set up a homepages, whereas only 56% of Koreans said "no." Asked if they are interested in buying a computer for the household, 46.6% of Japanese said "no," whereas 23.2% of Americans and 17.2% of Koreans said "no."

As the above findings indicate, Koreans have a greater interest in and more positive attitude toward the Internet as a communications tool. This enthusiasm is reflected in how the Korean media presents the Internet in society. With the financial crisis and presidential election in December of 1997, the Internet and CMC gained increased attention from two perspectives. The first is the use of the Internet as an economic tool to help revive the Korean economy. This includes reports on how SOHO (small office/home office) and venture capital companies use Web sites to attract business and on Web sites containing job placement information. Other reports show how Web sites are used effectively to promote exports or improve English-language skills, both of which are seen as critical to Korea's future. The second perspective is the effect of CMC on democracy in Korea. The media often carry reports of how chat rooms, BBS, and Web sites give people, usually marginalized groups in society, fora to express their views. Such reports became more common in 1998 as labor organizations set up Web sites to protest corporate restructuring, which threatened job security of blue- and white-collar workers in a broad range of industries.

Table 1
Nomura Survey on Public Attitudes toward the Internet and Computers

Q: Do computers and other information technology increase human communication?

	Japan	Korea	US
Yes	43.2%	75.4%	73.8%
No	56.4%	23.6%	25.0%

Q: Do you worry not being able to use a spreading new technology that spreads in society?

	Japan	Korea	US
Yes	59.6%	72.2%	70.8%
No	40.2%	27.8%	28.4%

Q: Do you want to let the world know of your existence and ideas?

	Japan	Korea	US
Yes	32.7%	59.0%	62.4%
No	66.8%	41.0%	34.6%

Q: Do you want to set up a personal homepage?

	Japan	Korea	US
Already	.7%	2.6%	4.0%
Yes	25.9%	40.4%	18.6%
No	71.2%	56.0%	70.0%

Q: Do you want to buy a computer for the household?

	Japan	Korea	US
Already	33.0%	47.6%	52.6%
Yes	19.3%	35.2%	21.6%
No	46.6%	17.2%	23.2%

Note: The above figures exclude "none-of-above" responses.

In my interview with him, Chang Yunhyŏn, the director of *Chŏpsok*, told me that early interest in CMC came from the desire for free expression that had been suppressed during the years of dictatorship that ended in 1987. According to Chang, the anonymity and spontaneity of chat rooms allowed participants free expression that was difficult in conventional media and even in face-to-face meetings. Chang himself was a student activist in the eighties, who, like many of his peers, became enchanted by CMC as it emerged in the early nineties. At times, chat-room participants suddenly agree to meet face-to-face at a set time and place, which suggests that chat-room interaction helps break the ice for face-to-face meetings (Kim and Ch'oe 1996). The Korean term for this is *pŏngaet'ing*, or "lightning meeting," consists of the *pŏngae* (lightning) and the last four letters of the English word "meeting." It also suggests that CMC augments, rather than supplants, face-to-face communication in Korea. In her ethnographic study of CMC in Korea, Kim (1996) found that most CMC users enjoy the anonymity of the medium, but do not expect it to replace face-to-face communication. Those who want more intimacy in the relationship try to meet face–to–face. And, as noted above, the Korean media often refer to Web pages and chat-room data as a gauge of economics, political, and social trends.

The Internet and CMC have a number of critics in Korea as well. During the 1997 presidential election campaign, the government decided to monitor domestic chat rooms for slanderous comments and negative campaigning. The prosecutor's office established a team of investigators to watch chat rooms. Yi Yonguk (1997) has argued that chat rooms and BBS postings encourage "cyber-sadism," which inhibits the development of democratic discourse in CMC. The prospect of government monitoring of chat rooms alarms Im (1996), who found that chat rooms air a much wider range of opinions than traditional media print and broadcast media. He fears that government monitoring will discourage free discussion and deprive CMC of its democratic potential. In 1998, sexual harassment and abusive language in chat rooms received increasing attention in the media. Two high-profile stories—one about couples using chat rooms to agree to swap partners and another about soliciting prostitution in chat rooms—were reported in all major newspapers. The issue has become serious enough to prompt the Korean government to require users of commercial services to use their real name and address when signing up for such services (Sŏk 1998). This would make it easier for companies and aggrieved individuals to monitor those persons who habitually abuse chat rooms. Social conservatives have argued that chat rooms have a negative effect on young people because they use bastardized slang and a variety of orthographic deviations (Kim 1997). Like pagers, chat rooms allow young people to escape parental supervision, which social conservatives argue encourages illicit contact with the opposite sex. The media are less critical of the Web, but run frequent articles on problems of teenagers accessing pornographic sites. One ironic story mentions the opening of a "swear room" under psychological supervision in several commercial on-line services. The "swear room" allows abusive language, but with the intent of giving users a space to vent their anger as a way to relieve stress. It thus tries to channel the urge to use abusive language in chat rooms into a form of on-line therapy (Kim 1998).

These observations, taken together, suggest a clear difference between Korean and Japanese cultures—with Korea comparatively more supportive of CMC than Japan. To explain this difference, the culture theory would be driven towards portraying Japan as a nation of "cyber-Luddites" or "technophobes." Evidence from other fields suggests the opposite. Japan was one of the first countries in the world to embrace high-speed trains. The fax machine caught on rapidly in Japan, and, according to the Nomura survey (Table 2), 20.2% of Japanese have a private fax machine, the highest of the four nations surveyed. Mobile phone usage in Japan is also high, with 35.7% of the respondents in the Nomura survey saying that they use a mobile phone. Another 25.3% of the respondents have a mobile phone in their household, but do not use it personally. Thus, 61% of the respondents said that they have a mobile phone in their household. This compares with 38% in the United States and 39.4% in Korea. Indeed, along with the Scandinavian countries, Hong Kong, and Singapore, Japan has one of the highest rates of mobile phone diffusion in the world. But if the culture theory seeks to explain more negative attitudes towards CMC in Japan as a function of an anti-technological "culture," these additional data directly contradict such an explanation. Indeed, this contrast between Japanese attitudes towards CMC and other technologies suggests that "culture" may be too general and vague as a concept to adequately explain such divergent attitudes within the same culture.

The second theory, the computer-literacy theory, focuses on how computer literacy and a variety of practical considerations, such as the cost of machinery and on-line time, affect attitudes toward the Internet and CMC in Japan and Korea. Though these issues have been discussed at length by proponents of the Internet, they have rarely been discussed in cross-cultural comparisons.

The spread of CMC faces several major obstacles in Japan, the first of which is the difficulty of inputting Japanese into a word processor. The problem is simple: the writing system requires two stages of inputting, which slows typing and makes it difficult for users to participate in chat rooms. Regardless of which system is used to input Japanese into the computer, users must press the space bar to bring up the desired combinations of Chinese characters, which are then entered into the text by pressing the enter key.

Table 2
Nomura Survey on Attitudes toward Non-Internet Communication Technologies

Fax Machine in Household	**Japan**	**Korea**	**US**
	20.2%	3.6%	17.0%
Mobile Phone Use	**Japan**	**Korea**	**US**
Use Personally	35.7%	16.2%	32.4%
Used in Household	25.3%	23.2%	5.6%

Note: The above figures exclude "none-of-above" responses. Figures for mobile phones exclude those who do not have one in their household.

This contrasts with English and Korean, both alphabet languages, in which the typed letters enter the text as they are typed. Results from the Nomura survey show that Japan has the lowest level of keyboard literacy of the four nations surveyed. According to the survey (Table 3), only 6.2% of Japanese said that they could type fast without looking at their fingers, whereas 29.8% of Americans could do so, and 16.8% of Koreans. The figures for those who could type fast while looking at their fingers was 17.5% in Japan, 24.6% in the United States, and 14.8% in Korea. Thus, only 23.7% of Japanese can type fast, whereas 54.4% of Americans could type fast and 31.6% of Koreans could type fast. The question on length of computer use also yields interesting results. As expected, Americans have the longest experience using computers: 42% have used computers for more than four years, whereas 20.8% in Japan, roughly the same number that can type fast, and 12.2% in Korea have used computer for more than four years. In Japan, 50.8% have never used computers, which contrasts greatly with the figure of 21.8% in the United States. In Korea, despite a lower per capita income than Japan, 49% of respondents said that they had never used a computer.

In a survey of Internet users in Korea, Yang (1996) found that perceptions of complexity, either of the computer itself or the Internet, had the greatest influence on user attitudes toward the Internet. Not surprisingly, users who viewed computers and the Internet as complicated had negative attitudes, whereas those who are familiar with computers and perceive going on-line as easy had positive feelings toward the Internet. In the future, voice recognition software may allow Japanese computer users to input words rapidly into the computer, but the small number of distinct syllables in the Japanese language makes it difficult for the computer to distinguish among the large number of words and syllables with the same pronunciation (Moffett 1998).

Another problem with the spread of the Internet and CMC in Japan is the cost of on-line services and local telephone calls. On-line services are not excessively expensive, but local telephone charges add up quickly. For many Japanese, going on-line adds to monthly expenses at a time of significant economic weakness. The perception of high cost makes it difficult for Japanese to go on-line without worrying about telephone charges (Kotkin 1999). Thus, of those who use a computer at home, only 34.6% of Japanese respondents in the Nomura survey said that they are connected to a network, whereas 63.4% in the United States, and 43.8% in Korea were connected to a network. One way around the problem would be to go on-line at work or school, but only 9.4% of Japanese respondents in the Nomura survey said that they were connected to a network. This contrasts with 27.8% in the United States and 10.2% in Korea. The low figure for on-line connections at work or school in Japan indicates that employers do not have the money to invest in computers or that they do not value networks in the workplace (Kunii 1998). In both countries, the size and financial stability of the organization concerned has a direct influence on the ability to invest in networks.

Table 3
Nomura Survey on Keyboard Literacy and Length of Computer Use

Typing Proficiency	**Japan**	**Korea**	**US**
Fast without Looking	6.2%	16.8%	29.8%
Fast but Look	17.5%	14.8%	24.6%
Slow and Look	39.2%	26.2%	31.8%
Barely Use	36.7%	42.2%	11.4%

Length of Computer Use	**Japan**	**Korea**	**US**
More than 2 Years	20.6%	12.2%	42.0%
2-4 Years	11.5%	12.4%	16.8%
Less than 2 Years	16.0%	26.4%	17.8%
Never Used	50.8%	49.0%	21.8%

Note: The above figures exclude "none-of-above" responses.

Conclusion: Toward a Synthesis

In comparing the culture and the computer-literacy theories of perceptions of CMC in Japan and Korea, the computer-literacy theory provides a better, if not simpler, explanation of the differences in both nations. The comparative slowness of word processing in Japanese, the cost of on-line time, and the affordability of competing communication technologies have inhibited the spread of CMC in Japanese society. In Korea, by contrast, word processing programs are easier to use, on-line time is economical compared with other communications technologies, and the media has paid greater attention to the role of CMC in society, all of which encourage the diffusion of the Internet and CMC in Korean society. Evidence from diffusion of high-speed trains and mobile phones from Japan shows that Japanese consumers are receptive to technologies when they are reasonably priced, convenient to use, and reach a critical mass of users. The perception of cost and difficulty among many Japanese takes the ease and spontaneity out of CMC, which makes it less appealing than the mobile phone. From the view of cost, convenience, and spontaneity, Japan can, at the risk of creating a new sound bite, be described as a mobile-phone society and Korea as a CMC society.

The computer-literacy theory fails to explain, however, explain the negative views of CMC in the Japanese media. As in many nations, the media in Japan reflect the views of the ruling establishment, which failed to realize the potential of the Internet in the early nineties. Without the tacit "guidance" from the ruling elite in Tokyo, the media is free to pick up on public misgivings over the computer as a tool for communication. Korean policy makers have not promoted the Internet and CMC as aggressively as Singapore, but Korea's dependence on exports makes it imperative that it reach out to overseas markets. Interest in learning English and other foreign languages is higher in Korea than in Japan. As a percentage of the population, more Koreans live and study overseas than do Japanese. The socio-economic atmosphere in Korea contributed to a positive impression of CMC as a way to cut across time and geography to reach customers, relatives, and friends around the world. Media and public enthusiasm for CMC in Korea will ensure continued diffusion, despite reports of hacking and abusive language in chat rooms. The recent economic crisis and change of government will encourage more rapid diffusion, particularly in companies and schools. Reports of "virtual universities" formed by a consortium of a number of universities are a sign that this is already happening (*Chosŏn Ilbo* 1998). As another example, an elementary school in Taegu, a large city in southeastern Korea, has started computer classes for parents so they can assist their children in using computers at home (Oh 1998).

In Japan, however, diffusion will proceed slowly with worries over telephone charges and the effect of CMC on communication patterns in Japan. The establishment is torn between trying to reform the post-war system and promoting dramatic far-reaching reforms. Should the establishment embrace the Internet and CMC and institute policies that foster diffusion, media and public perceptions will become decidedly positive, and rapid diffusion will follow. The following story from another leading Korean newspaper, the *Tong-A Ilbo*, tells of an Internet policy success in Japan (Yi Munung 1997). In the village of Yamada in the Toyama Prefecture, the prefectural government gave 325 of 548 villagers a PC with videophone functions. All citizens were offered computers, but some refused for personal and health reasons (age, eyesight, etc.). Every citizen has an Internet ID so that they can send e-mail or communicate by videophone. Village meetings are held through Internet conferencing and the village homepage (<http://www.vill.yamada.toyama.jp>) has links to villagers with individual homepages. The village and individual homepages promote local agricultural products and recreational facilities. The success of this project shows that CMC will spread rapidly in Japan or anywhere else when it is perceived as convenient, economical, and as enhancing existing patterns of communication.

Notes

I would like to that the International Studies Research Institute at Kumamoto Gakuen University for providing a research grant to visit Korea in February 1998.

1. The data for this study were taken from a variety of Japanese and Korean Web pages, newspaper and magazine articles, and Master's degree theses. I will also refer to an interview with Chang Yunhyŏn, director of *Chŏpsok*, a popular Korean film about a chat-room romance. Largely, my sources reflect what was available on the Web from my office in Japan and references that I could gather during a weeklong research visit to Korea in February 1998.

References

Asahi Shimbun. 1996. Renai no sokudo wa soto no sekai no nibai, sambai ("The Speed of Falling in Love Is Two to Three Times as Fast as the Real World"), 26 September. (Obtained on-line from *asahi.com*, <http://www.asahi.com>).

Asahi Shimbun. 1997. Pasokon tsūshin de chūshōsareta' ("I was Wounded On-line"), 27 May. (Obtained on-line from *asahi.com*. <http://www.asahi.com>).

Asiaweek. 1998. 11 December, 74.

Bijutsu Techō. 1997. PostPet' tte nanda!? ("What on Earth is 'PostPet'!?") 741: 18–25.

Chosŏn Ilbo. 1997. Yŏnghwa 'Chŏpsok' kwangaek ilbaengman myŏng tolp'a ("The Film *Chŏpsok* Attracts over a Million Viewers"), 21 October. (Obtained on-line from *Digital Chosŏn*. <http://www.chosun.co.kr>).

Chosŏn Ilbo. 1998. Kasang taehak' p'ŭrogŭraem shibŏm unyŏng kigwan tasŏt kot sŏnjŏng ("Five Institutions Selected to Pilot Test the 'Virtual University' Program"), 13 February. (Obtained on-line from *Digital Chosŏn*. <http://www.chosun.co.kr>)

Courrier International. 1998. Parlez Used zhongwen? 16–22 July, 27.

Connery, Brian A. 1997. "IMHO: Authority and Egalitarian Rhetoric in the Virtual Coffeehouse." In *Internet Culture*, ed. David Porter, 161–79. New York: Routledge.

Fouser, Robert. 1998. "Naked Lunch." *World Art* 17: 30–34.

Hong Sŏkp'yo. 1998. Personal communication, 21 October.

Im, Yŏnghwan. 1996. PC t'ongshin ŭi kongnonjang yŏkhal e kwanhan yŏn'gu ("A Study on the Function of CMC in the Public Sphere"). Master's Thesis, Hanyang University.

Itami, Kazuhiro. 1997. E-mēru ga hirogeru 'hirei' no otoshi ana ("The Pitfall of Rudeness Spread by E-mail"). *AERA*, 15 September, 58–59.

Kim, Kyŏnghwa and Ch'oe Inch'ŏl. 1996. PC t'ongshin t'onghan chŭkhŭng deit'ŭ: 'pŏngaet'ing' kalsurok ingi ("Exciting Dates through CMC: 'Lightning Meetings' Gain Popularity"). *Han'guk Ilbo*, 23.

Kim, Chŏnghŭi. 1996. Taein k'ŏmyunik'eisyŏn ch'aenŏl rosoŭi k'ŏmp'yut'ŏ maegae k'ŏmyunik'eisyŏn ("CMC as an Interpersonal Communication Channel"). Master's Thesis, Seoul National University.

Kim, T'aeik. 1997. PC t'ongshin i kugŏ oyŏm shik'inda ("CMC is polluting the Korean Language"). *Chosŏn Ilbo*, 14 December, 17.

Kim, Chaesŏp. 1998. Yojŭm sesang yok ina shilk'ŏt: PC t'ongshin 'yokpang' ttŏtta ("Swear at the World: 'Swear Rooms' Debut On-line"). *Hangyŏrae Shinmun*, 8 August: 8.

Kotkin, Joel. 1999. "New Home for a Lost Generation of Innovators." *The New York Times*, 28 March. (Obtained on-line from <http://222.nyt.com>).

Kunii, Irene. 1998. "Will Technology Leave Japan Behind?" *Business Week*, 31 August, 124–26.

Moffett, Sebastian. 1998. "Kissing Kanji Goodbye?" *Business Week*, 23 February, 56.

Nomura Research Institute. 1998. Jōhō tsūshin riyō ni kansuru kokusai hikaku chōsa o jisse ("Results of a Comparative International Survey on Information Technology Usage"), 12 February. (Obtained on-line from <http://www.nri.co.jp/news/980212/index.html>).

Oh, Taeyŏng. 1998. Taegu Hwanggŭm Ch'odŭnghakkyo pumo ga hamkke hanŭn 'saibŏ hakkyo' ("Hwanggŭm Elementary School in Taegu: A 'Cyber-School' that Includes Parents"). *Chosŏn Ilbo*, 9 July. (Obtained on-line from *Digital Chosŏn*. <http://www.chosun.co.kr>).

Rheingold, Howard. 1993. *The Virtual Community: Homesteading on the Electronic Frontier*. Reading, MA: Addison-Wesley.

Sŏk, Chonghun. 1998. PC shilmyŏngje: i dal chung shihaeng ("PC Real Name System to Be Adopted this Month"). *Chosŏn Ilbo*, 7 December. (Obtained on-line from *Digital Chosŏn*. <http://www.chosun.co.kr>).

Stoll, Clifford. 1996. *Silicon Snake Oil: Second Thoughts on the Information Highway*. New York: Anchor Books.

Takenaka, Hideo. 1998. Dennō famiri 'genjitsu' to 'kasyō' no aida de ("The Digital Family: Between Reality and Virtuality"). *Yomiuri Shimbun*, 9 January. (Obtained on-line at <http://www.yomiuri.co.jp/etc/nippon.htm>).

Ttanji Ilbo. 1998. *Ttanji Ilbo 1*. Seoul: Chajak Namu. (Also see on-line edition at <http://ddanji.netsgo.com>).

Turkle, Sherry. 1996. "Virtuality and Its Discontents: Searching for Community in Cyberspace." *The American Prospect* 24(1): 50–57.

Watanabe, Yasushi. 1997. 'kōshin' kara 'senshin' e: Saibā Ajia no jōhō infura ("From 'Backward' to 'Advanced': Information Infrastructure in Cyberasia"). *InterCommunication* 19: 104–109.

Yang, Ch'anil. 1996. Int'ŏnet ŭi hwaksan kwajŏng e yŏnghyang ŭl michinŭn yoin e kwanhan yŏn'gu ("A Study on Factors Influencing the Diffusion of the Internet"). Master's Thesis, Chungang University.

Yi, Munung. 1997. Ilbon ŭl nollage han int'ŏnet maŭl Yamada mura ("The Internet Village that Surprised Japan"). *Tong-A Ilbo*, 11 November, 21.

Yi, Yonguk. 1997. Kaeship'an munhwa nŭn chŏnja minjujŭi ŭi sŏngp'ae karŭm hanŭn shigŭmsŏk ("Public On-line Postings are the Testing Ground for Electronic Democracy"). *Munhwa Yesul*, 218: 16–20.

III. Cultural Collisions and Creative Interferences on the (Silk) Road to the Global Village: India and Thailand

Language, Power, and Software

Kenneth Keniston

In discussions of the impact of "The Information Age," the role of language in computing is rarely mentioned. Hundreds of books have analyzed the digital age, the networked society, the cyberworld, computer-mediated communications (CMC), the impact of the new electronic media with hardly a word about the central importance of language in the Information Age.

The goal of this paper is to give language—by which I mean the language in which computing is done and in which computer-mediated communication occurs—a key place in discussions of the impact of computation and computer-mediated communications. I will argue that the language in which computing takes place is a critical variable in determining who benefits, who loses, who gains, who is excluded, who is included—in short, how the Information Age impacts the peoples and the cultures of the world. In other words, I will stress the relationship of language to power, wealth, privilege, and access to desired resources.

Localization and Language

Although the ultimate "language" of the computer consists of digital zeroes and ones, the language of users, including programmers, is and must be one of the thousands of existing languages of the world. In fact, virtually all programming languages, all operating systems, and most applications are written originally in English, making language a "non-issue" for the approximately seven percent of the world's population that speaks, reads, and writes fluent English.

Since all major operating systems and applications are written in English (with the exception of the systems written for the German firm, SAP, which specializes in accounting software), use by non-English speakers requires localization. Localization entails adapting software written in one language for members of one culture to another language for members of another culture. It is sometimes thought to be simply a matter of translation. In fact, it involves not only translation of individual words, but deeper modifications of computer codes involving scrolling patterns, character sets, box sizes, dates, dictionary search patterns, icons, etc. Arabic and Hebrew scroll from right to left, unlike the North European languages. Russian, Greek, Persian and Hindi involve non-Roman character sets. Ideographic, non-phonetic written languages like Chinese and Japanese involve tens of thousands of distinct characters.

Translation alone is an exceedingly complex part of localization. Ideally, it is a multistage process involving initial translation, followed by "back-translation" into the original language, comparison of the back-translated text with the original, adjustment of the translation as necessary, and incorporation of the now corrected translation into the final localized program. The cost per word thus translated has been estimated as approximately one dollar. Given that large programs like operating systems or office suites may contain tens of thousands of pages of text, localization even at the level of translation is both complex and expensive.

But localization involves more than simple translation. Scrolling patterns, character sets, box sizes, dates, and icons must be adapted to the new language and the culture in which it is spoken. As one observer has noted with regard to computer icons, there is no gesture of the human hand which is not obscene in some language. As others have noted, the color red, which indicates "stop" or "danger" in the US, may indicate life or hope in another culture. Dictionary search patterns in a language like Finnish, which is highly inflected, require searching out the root verb from a word which may contain as prefixes and suffixes what in English would be the balance of an entire complex sentence.[1]

Moreover, localization is a worldwide business of growing economic importance. The industry association, the Localization Industry Standards Association (LISA), in Geneva holds periodic meetings of localizers and publishes a newsletter (<http://www.LISA.unige.ch>). Every major software firm has a localization division, and many attribute large parts of their sales not to the original English language version, but to localized versions sold in other countries. More than half of Microsoft sales are outside the United States—although not necessarily in languages other than English. As an industry, the localization industry is highly diverse and not geographically concentrated.

Other than the localization divisions of major software firms, there are literally hundreds of firms, scattered throughout the world depending on the linguistic area, which "specialize" in localization, often on subcontract from major software producers. Indeed, the software giants of the US often turn to small partners abroad to localize, or to test localized versions of, their major packages. To my knowledge there is no study of the history and organization of the localization industry.

Localization is ordinarily seen as primarily a technical task. The localizer must not only be an experienced code writer, but must have a thorough knowledge of two languages, and ideally, of two cultures. Even localization from one North European language to another (e.g., from English to Spanish) requires good coding ability together with a knowledge of the subtleties of both languages.

"Localization" is intimately linked to another issue, commonly termed "standardization of code." To understand the importance of standardization requires analyzing how computers interpret letters—the letters, say, of standard English. Since computers can deal only with digital numbers, American computer coders decided early on that the letters of the English language (along with numbers, punctuation marks, etc.) would be mapped onto an eight-bit grid (which contained 256 theoretical possibilities). The standard known as ASCII (American Standard Code for the Interpretation of Information) assigns to each letter, number, and punctuation mark a specific numbered place among the 256 possible places. Thus, for example, the letter "lower case a" might be assigned location number 27, "lower case b," 28, etc. Computers (which communicate only in binary numbers) have established a convention that each 8-bit word ("byte") at some point contains an alphanumerical symbol. The decoding software "reads" from a positive sign in location 27 in a 8-bit "table" the letter "c," for example, location 27 representing the letter "c," which is then placed as a "c" on the screen, stored as a character, added it to another word, printed as a "c," etc. Communication between two computers is possible when they all use the same standardized code, such as ASCII. ASCII emerged to solve the problem of lack of standardization. In an earlier period, each software manufacturer devised his or her own proprietary system for alphanumeric coding. Thus, one system's "a" may have been location 27, while another's was location 203. Cross-platform intelligibility was impossible; each proprietary system required mastery of its own internal code; communication between two computers using different codes was impossible (or required complex transliteration programs). To solve this Tower of Babel problem, ASCII was developed and little by little imposed by its success on virtually all American software writers, and then, with modifications, on other languages whose characters could be adapted to the eight-bit ASCII system. With modifications, ASCII, or a comparable eight-bit (one byte) system, has proved adaptable to most languages except the ideographic languages like Chinese, which require tens of thousands of characters. For them, two-byte codes are necessary, involving 256^2 (65,536) possibilities. The emerging standard called Unicode, which aims at including all human languages, is a two-byte system.

But localization—whether it occurs, how it occurs, and how well and deeply it is done—is also an area where technology meets politics and culture in ways that I will emphasize in this paper. Elsewhere I have pointed to the ways that implicitly embedded cultural assumptions of the original language (almost always English) may (even in well-localized software) be perceived as alien, hostile, or unintelligible to users in another culture (Keniston 1997). Here I will focus on the prior question of whether or not localized software exists at all.

Localization, or more generally language, has rarely been treated as an important topic in the literature on the impacts of the so-called Computer Age. But both individuals and governments have been acutely aware of this problem. The Indian high school student in Delhi with a perfect knowledge of Hindi but a less than perfect knowledge of English confronts the issue of localization daily when he struggles with the "help" menus of his Windows 98 operating system—in English. The government of the tiny island republic of Iceland (population 500,000) confronts the issue of localization directly when it pleads with Microsoft to develop an Icelandic version of Microsoft's operating systems on the grounds that in its absence, young Icelanders are losing fluency in their traditional language. Of all nations, France has been perhaps the most vigorous in insisting on localization. A former French foreign minister termed the effort to preserve the hegemony of French against English "a worldwide struggle," "which we, the French, are the first to appreciate." Allying themselves with French-speaking Canadians and French speakers in so-called "Francophonic Africa," the French have made systematic efforts to suppress the use of English and insist on French. Software imported to France and Web sites developed in that country must use French as a matter of law. For the French, the enemy is the "Anglophonic tide." These French concerns are shared, though often less articulately and less overtly, in other parts of the world. A senior German telecom official recently commented, off the record, that German concerns over the hegemony of English in the computer world were almost as intense as those of the French. "But," he added, "we let the French do the talking for us."

More important, worries about the "Anglophonic tide" in software merge with deeper worries about the power of so-called "Anglo-Saxon culture" on local values. What is the impact on villagers in African hamlets when satellite television permits them to see "Dallas," even if dubbed in Hausa, Igbo, or Swahili? How do Indian villagers react to Indian MTV, brought to them via satellite courtesy of Star TV, and MC'd in English by a laid back young Indian with an American accent? How does the spread of computers and computer-mediated communication (Internet, Web) influence existing inequalities of power within each society? How does it influence the gap between the rich societies of the North and the poor societies of the South? And does the dominance of English as the language of computation, Internet, and the World Wide Web contribute to undermining the vitality and richness of ancient, non-Anglo-Saxon cultures, especially in Africa and Asia?

These questions are too rarely asked, perhaps because they have no simple answers. Yet if we agree that the new electronic technologies are the most innovative and powerful technologies of the new millenium, then these questions, however difficult, must be asked. How do the new electronic technologies affect existing inequalities within and between nations? How do they impact the cultural diversity of the world?

Information Technology in South Asia

The seven nations of South Asia are in some respects unique, in some respects important in themselves, and in some respects illustrative of problems faced by many other regions. The basic facts about South Asia are well known. Approximately one fourth of the world's population (1.2–1.3 billion persons) lives in the seven nations of India, Pakistan, Bangladesh, Sri Lanka, Nepal, Bhutan, and the Maldives. An estimated 5% of this population speaks good English, giving the subcontinent the second largest English-speaking population in the world, ahead of Great Britain and led only by the United States. English language fiction today is strongly influenced, indeed perhaps dominated, by writers of South Asian origin.[2] Indeed, the articulateness of educated South Asians in English is legendary. For the English-speaking segment of the South Asian population, computing, almost entirely founded on the English language, presents no problems whatsoever, nor does computer-mediated communication (e-mail, Internet, Web) in English.

There are, however, approximately 1.2 billion people in the Asian subcontinent who do not speak (or more important from the point of view of computation, read and write) good English. To begin with, approximately half of the population of the subcontinent is not literate at all. Equally important, most of the vast literate population of the region is literate in some language and script other than English—or for that matter other than French, German, Spanish, etc., languages for which localized software is available for all major operating systems and many important applications.

South Asia contains some of the world's largest linguistic groups: for example, Hindi with an estimated four hundred million speakers (approximately the population of the European Union), Bengali with approximately two hundred million, and languages like Telegu with eighty million (about equal to the population of Germany.)[3] There are literally dozens of languages with more than a million speakers in South Asia. India alone recognizes eighteen official languages. Most of these languages have a unique script, and most have important literary traditions, both oral and written, that go back millenia. Some languages are cognate: for example, Urdu and Hindi both derive from the Hindustani of the Northern Plains, the one Persianized and the other Sanskritized in accordance with the cultural and political dictates of their respective speakers and nations.

In India today, major linguistic conflicts are largely absent. The initial plan to impose Hindi as the national link language has been repeatedly abandoned in the face of resistance from non-Hindi-speaking Indians, especially in the Southern states. The Indian states have been organized along linguistic lines, while English is accepted as the *lingua franca* of the national legislature, the higher civil service, the higher (national) courts, most highly educated people, and most national and multi-national businesses.[4] But in Pakistan linguistic issues were central in the split between East and West Pakistan (what is now Bangladesh), and conflict over the role of Urdu, Punjabi, Sindhi, and other languages continues in today's

Pakistan. In Sri Lanka, the Sinhala- and Tamil-speaking populations have deep and destructive conflicts. So any simple generalization about the role of language in South Asia fails. In India, language is largely a non-issue in the political sense; in other nations, it is a cause or symbol of violent political polarizations.

One fact is constant, however. Throughout the entire subcontinent, English is the language of wealth, privilege, and power. For this reason, in Karachi, Dakha, Delhi, Colombo, and Katmandhu, parents who can afford it commonly seek English-language instruction for their children, aspiring to fluency in English at least as a second language in order to open to their children access to positions of responsibility, wealth, privilege, and power in their own societies and abroad. An Indian colleague tells of Hindu-nationalist villages in the most fundamentalist areas of India where every fourth shop on the streets offers English language instruction.

That English is the language of power, wealth, prestige, and preferment in South Asia is no accident. As many have documented, in the 1830s the English policy-maker Macauley laid down the rules that guided English colonial educational work in India (and elsewhere) from the start. His goal was to use the English language, and to import English pedagogic methods and content in order to create a leadership group of "brown skinned Englishmen," infused with English cultural values and loyal to the Empire. For more than a century, in India as well as in English colonies in Africa, Singapore, Malaysia, Hong Kong, and elsewhere, this plan guided British colonial linguistic policy.

Lord Macauley was a complex figure, an imperialist to be sure, but one who foresaw the day when India would claim independence as what he termed the "proudest day" in Great Britain's history.[5] Moreover, in his belief that learning a language meant acquiring a culture, he anticipated the thinking of many modern applied linguists. One need not believe that language is reality in order to acknowledge that each language makes it easy to say some things, difficult to say others, and impossible to say still others. In short, language shapes, organizes and structures what we can communicate, how we think, and what we experience.[6] I recently worked with an MIT student brought up in Korea who was losing his facility with the Korean language. I expressed my regret and urged him to keep up his fluency. He commented with perception, "It doesn't really matter, because I can still think Korean." In other words, he was asserting that knowing a language entails knowing a way of organizing reality.

If Macauley's policy succeeded linguistically at least with Indian elites, it failed dramatically in other ways. As the independence movement of India and other former British colonies showed, that policy failed to imbue in the population of South Asia, and even in English-speaking elites, an undying love for British rule and Empire. Politically, Macauley's policy was a complete failure, even if culturally it was partially successful. Men like Gandhi and Nehru in India, or Jinnah in Pakistan, attacked the British raj in exquisite English, which they had often learned in English public schools and universities. Indeed, some have even claimed that "Anglo-Saxon" values of fair play, equality, the rule of law, and the dignity of all human beings paradoxically helped inspire the movements of independence of the former British colonies.

Studies of the elites of South Asia are rare and incomplete. Clearly, these elites differ from nation to nation, from region to region, from city to city. The Urdu-speaking elite in Pakistan that resulted from Partition differs in important respects from the business elite of Bombay or the political elites of Delhi. Moreover, with dramatic changes underway in the subcontinent, generalizations valid a decade ago may be invalid today. Witness, for example, the rise of a new younger generation of entrepreneurs in India, fueled by the progressive "liberalization" of the economy. Witness, too, the emergence of an elite group of the "captains of the software industry," today India's largest source of export earnings.

But whatever the characteristics of elites in South Asian cities and nations, they tend to have one common characteristic. For membership in South Asian elites, English is not only useful, but it is virtually the only privileged route to power, the only reliable key to any reasonable hope of wealth, preferment and influence. In South Asia, as in few other regions of the world, language and power are fused. To be sure, English plays a similar role in the distribution of wealth, power and influence in other former British colonies in Africa and Southeast Asia. Moreover, throughout the world, English is today the preferred language of commerce and science, a fact almost as true in North Europe as it is in South Asia. In South Asia, however, the fusion of language and power is almost total.

What makes this relevant for computation and the impact of the Information Age in South Asia, and what differentiates South Asia from many other parts of the world, is the nearly complete absence of localized software in any of the traditional languages of this vast and populous region. Efforts have been made to change this situation; many schemes for localizing programs, operating systems, and applications to vernacular languages exist; many creative people are working on this problem. But the fact remains that, as of early 1999, none of these "solutions" has achieved any widespread acceptance. There are more plans than achievements; the policies of the Indian Government vis-à-vis localization remain complex and confused. Despite multiple proclamations on the part of both public and private groups that they have achieved a solution to the localization problem, either these solutions do not work or they are not widely adopted.

The result is that South Asia—with its vast population, its enormous economic potential, its multiple ancient cultures and literatures, and the world's largest, rapidly growing middle-class—almost completely lacks readily available, affordable, usable vernacular software. To put it bluntly and perhaps to overstate the point, unless an Indian reads, speaks, and writes good English, she cannot use a computer, she cannot use e-mail, she cannot access the Web. Despite the valiant efforts of many who have tried to change the situation, English is necessary.

Why Is There No Local Language Software?

Given that South Asia possesses almost a quarter of the world's population, we need to ask why there is no effective and diffused localized software. An answer requires examining different levels of the problem.

First is the question of why the efforts of software companies in this area have been so meager or so ineffective. At the governmental level, India has promoted two distinct groups concerned with local language software, the National Centre for Software Technology (NCST) in Mumbai, and the Centre for Development of Advanced Computing (CDAC) in Poona. Each has followed a different path toward localization, with CDAC the first to market. CDAC's solutions were initially based on hardware modifications (the so-called GIST card), and its word-processing software was seen by some users as inadequate and antiquated. Furthermore, CDAC, although a government agency, initially sold its local-language software, warts and all, for prices that drove away potential purchasers of lesser means. NSCT, which currently works with Microsoft on developing Indian language fonts, has developed alternative means of coding Indian languages, which many viewed as more likely to prevail than those promoted by CDAC. In Delhi, many agencies were directly or indirectly involved with setting policies that affect Indian language computing, including a special Government of India agency to promote the use of Indian languages, the Department of Telecommunications, and the Regulatory Authority of India. Competition or non-communication between these groups often resulted in conflicting rules or incompatible standards. Early on, of course, Indian computer scientists fully recognized the need for standardization of the major Indian languages and developed a coding system termed ISCII. ISCII is currently seen as more or less adequate for the northern Indian languages (which are based on Sanskrit and of Indo-European origin), but it is criticized as inadequate for the southern (Dravidian) languages. Indeed, a recent meeting of Tamil-speakers from India and other countries rejected the use of IISCI in favor of another, proprietary code.[7]

At the corporate level, too, efforts have also been ineffectual or non-existent. Microsoft, which controls 95% of the operating system business in India, has a number of collaborations like that with NCST, to develop Indian language capabilities for its programs. Microsoft has announced publicly that the next version of Windows NT (Windows 2000) will contain "locale coding" ability for two Indian languages, probably Hindi and Tamil. But "locale coding" is not localization. Rather, it involves the capacity to use a basic English language program such as Word in order to input and print another language. Thus, for example, locale coding for Hindi entails a system of keyboard mapping such that the individual can input Hindi characters [either phonetically or through direct (stick-on) keys], an internal software architecture that recognizes, interprets, and organizes these characters for output, and a set of fonts for monitor display and printing utilizing Hindi (Devanagari) characters. Although it is a step in the direction of localization, locale coding for Hindi nonetheless requires the ability to operate Windows and Word in English, and, in the case of keyboard mapping that uses the Roman keyboard phonetically, knowledge of the Romanized phonetic versions of Hindi words. Although it permits English-speakers to use the computer as something like a Hindi typewriter, it presupposes an advanced level of English.

Other multinationals and Indian firms have taken steps in the direction of localization. The MacIntosh interface lends itself to localization, and Apple has been a pioneer in localizing to Indian languages. The pity is that MacIntoshes are virtually unheard of in India, where they have less than one percent of the market. IBM announced in 1997 a Hindi version of MS-DOS. The pity here is that MS-DOS has not been used as a programming language or operating system for many years in most nations. Modular Technologies in Poona has a series of innovative products that permit the use of several Indian languages. BharatBhasha, organized by the brilliant computer scientist Harsh Kumar, has made available as freeware an overlay for Microsoft operating systems that permits their use in a number of Indian languages. The ironic pity here is that since BharatBhasha is freeware, distributors have no financial motivation to circulate it, and its use is still limited. Finally, with the advent of Internet, literally dozens of "Internet solutions" for Indian languages are available on the Web for free. The pity there, however, is that most of these solutions are mutually incompatible: if you have Hindi system A and I have Hindi system B, their coding of Hindi characters is different and we cannot communicate with each other.

In short, despite valiant and brilliant efforts to develop local language software, their impact has been restricted. Of the major players, only Microsoft and the Government of India have the clout to create universally-shared standards for the Indian languages and to build the localized software that would use them. Microsoft has chosen to focus its efforts on distributing English language software to the potentially large English-speaking Indian market, and, as noted, on developing locale coding for two or more Indian languages. The Government of India's efforts have been dispersed in a variety of activities, often brilliant but together not effective in creating widely-used local language software.

The fact thus remains that the Gujerati merchant who would like to computerize his operation so that he does not have to stay up until midnight balancing his books can find no small business applications except in English. The grandson from Delhi studying in London who would like to send e-mail to his Hindi-speaking grandmother in Delhi must do so in English or not at all. The dynamic major Indian software firms, oriented toward exports and services, have shown little interest in localization. The creative work done by many Indian individuals and groups has so far not produced effective applications in the major Indian languages. Even with regard to on-line Indian newspapers, most of which are not in English, the lack of standardization is consequential. Since few newspapers share the same coding of, for example, Hindi, for each Hindi newspaper on the Web, the Web user must download the separate set of proprietary fonts used by that newspaper.

Computers, Power, and Global Monoculture

In the spring of 1998, US President William Clinton spoke at the Massachusetts Institute of Technology on the Information Age. He devoted the first part of his talk to the wonders and potentials of the new digital technology. He stressed how it opens doors, provides access to information, facilitates communication, and aids commerce and education.

But in the second half of his talk, President Clinton pointed out that computers and computer-mediated communication also have the potential to widen the gap between the computer "haves" and the computer "have-nots." As the haves increase knowledge, power, and access to resources, the gap between them and those who are "computer-deprived" grows. In the United States, where at present almost half of all households have computers, and of them about half are connected to the Internet and the Web, those who benefit most from the Computer Age are those who already possess the greatest resources, political power and wealth.[8] The "information-deprived" are those who are already deprived in many other ways as well. Clinton ended his address by suggesting that market forces alone would not be enough to remedy this gap: both public action and private commitment are required to make the benefits of the Computer Age accessible to all.

In countless respects, the situation in South Asia is different from that in the United States. But in one respect it is the same: in both parts of the world, access to computers is empowering, and inability to access computers perpetuates deprivation, exclusion, and poverty. Indeed, as a general maxim in the history of technology, new technologies are appropriated by those who have power, and deliberately or not, these technologies serve initially to extend the power of those who already have power. In this regard, electronic technologies simply follow an historic rule.

But in South Asia, this universal problem is compounded by the overlap of power and language. Members of Indian elites are almost invariably English-speaking; India's vast population of peasants, tribals, scheduled, and backward castes—the excluded and deprived (many of them illiterate)—rarely know any but a few words of English. This convergence of language and power in India means that in special ways, the Information Age perpetuates the powers of the English-speaking elite; it widens the already large gap between those who now have both power and English, and the nineteen out of twenty Indians who have neither. No one planned it this way, but the dominance of English as a computer language helps perpetuate existing inequities in South Asia.

The second important issue stemming from the importance of English in computers in South Asia is the issue of cultural diversity versus an emerging global monoculture. The political scientist Benjamin Barber has recently argued that world culture is increasingly polarized around two extremes (1995). The first is what he calls "McWorld": the cosmopolitan, international, consumerist, multinationalized, advertising-based culture of cable TV, popular magazines, Hollywood films—a culture which aims at universal accessibility, in which billions watch the same World Cup finals, a culture where MTV (translated), dramatizations of the lives of imaginary American millionaires, CNN, and films like *Titanic* dominate and flatten local cultures, producing a thin but powerful layer of consumerist, advertiser-driven, entertainment-based, and perhaps in the last analysis, American-influenced culture with great popular (if lowest denominator) appeal, backed by enormous financial and technological resources. It almost goes without saying that this culture is, in origins and assumptions, predominantly English-speaking. Its centers are the US, Britain, Australia, English-speaking Canada, and English speakers in nations and city-states like Hong Kong, Singapore, South Africa . . . and India.

In defining the power of this global monoculture, computers, Internet, and the Web play a small but growing role. In South Asia, countless million Indians have access to cable television, while three or four million at most have computers, and of them, perhaps ten percent have access to Internet and the Web. The driving forces of Anglo-Saxon global monoculture are still television and film. But the dominance of English in computation is part of this broader picture, and its importance is likely to increase in the years ahead. With the liberalization of Internet service providers in India, with efforts to lower the costs of local telephone connections, and with the plummeting price of computers, more and more Indians are likely to join the "wired" world. Rates of Internet growth are higher in South Asia than in most English-speaking nations, although the starting base is low and there are virtually no non-English Web sites or Internet hosts in these nations. At the same time, however, the dominance of English as defining the wired world remains intractable: indeed, an article in *Salon*, the on-line Apple magazine, several years ago spoke of "the English speaking Web" (Brake 1996). While some counterexamples exist (Hongladarom, this volume), the world of computers and computer-mediated communication must be counted almost exclusively as McWorld, not of cultural local diversity.

The Japanese sociologist Toru Nishigaki of the University of Tokyo sees a global Anglo-Saxon monoculture ultimately based on the power of American entertainment and American values as threatening to marginalize all local cultures (see <http://lpe.iss.u-tokyo.ac.jp/>). He notes that a Japanese businessman who is fluent in Chinese and wishes to communicate with a Chinese partner must, today, first translate his thoughts into English, communicate them in English via Internet to his Chinese partner, who must in turn retranslate them into Chinese. Equally emblematic of the power of American culture is the power of American technology. Given the low cost and effectiveness of American communication technologies, it often proves less costly and more efficient to send a message from Bombay to Calcutta via satellite through the United States than directly across India.

At the opposite pole from McWorld, Barber sees the ugly side of fundamentalism, which he terms "Jihad." He persuasively claims that one reaction against the cosmopolitan, internationalist, multinational- and consumer-driven culture of McWorld is a return to the allegedly fundamental truths and varieties of an ancient culture. War is justified as an emblem of identity, an expression of community, an end in itself. "Even when there is no shooting war, there is fractiousness, secession, and the quest for ever smaller communities" (Barber 1992, 60). At worst, this return is exclusionary and even, as in the case of Jihad, may require holy wars against the impure. Jihad imagines a world of cultural and/or ethnic purity from which foreign, cosmopolitan, and alien influences have been eliminated, and in which an imagined ancient culture thrives, isolated from the rest of the corrupt and corrupting world. It is the world of "ethnic cleansing."

What Barber discusses as Jihad, however, also in his view has a different and friendlier face, namely that of cultural diversity. And in no part of the world is cultural diversity more manifest than in South Asia, and especially in India. Communal, religious, and ethnic tensions indeed exist and led, at the moment of Independence, to the tragedies of Partition and to repeated episodes of communal violence. Yet the fact is that India is the second largest Islamic nation in the world, with more than 170 million Muslims living—99.99% of the time—in relative harmony with their Hindu neighbors. India is also the most multilingual and multicultural major nation on earth. Linguistic and cultural divides have torn apart or threatened to dismember nations like the former Soviet Union, Yugoslavia, Czechoslovakia and Canada, but in India they have by and large been managed harmoniously. No subcontinent in the world possesses so rich and diverse a set of cultures as South Asia.

The preservation of cultural diversity in the world, and in South Asia in particular, is a high value, perhaps on a par with the reduction of inequity and the promotion of political freedom. Cultural diversity can, of course, be perverted into reactionary fundamentalism. But this is most likely when local cultures are deprecated, spurned, marginalized, viewed as inadequate, and when their members experience exclusion, condescension, or discrimination because of their membership in the culture. There is, then, every reason to value local cultures and to seek to make information technology a medium for their preservation and enhancement, not an instrument in their marginalization.

Given strong arguments that would support the creation of robust local language software in the major languages of South Asia, we need to ask why so relatively little has been done, despite the many voices raised to encourage vernacular computing. After all, the World Bank estimates that in the year 2020, India will have the world's fourth largest economy and the world's largest population. It is, of course, a poor nation at present, but it is also a thriving democracy, a nation with five hundred million literate men and women, a nation with a rapidly growing middle class, and a nation which is, as Bill Gates put it, a "rising software superpower." India has twice as many university graduates as the People's Republic of China, although much higher illiteracy rates. In short, India, and South Asia more generally, is a region where one could anticipate a rapidly growing market for local language software in the decades ahead. Yet as I have noted, few are responding to this emerging market. Instead, what appears to be a "Tower of Cyber-Babel" may be emerging with regard to Internet communication, and vernacular software remains, at best, a niche market.

Why So Little Local Language Software?

Among the reasons for the relative absence of local language software, economic factors surely play a key role. Indeed, it is often said that were there a market, localized software would simply appear. Indians as a group are poor; telephone penetration is low (and therefore Internet penetration is necessarily low). It can be argued that, given the fusion of language, wealth and power in India, there is simply no market (and perhaps no need) for software in any language other than English. Asked about localization to Indian languages, international software firms sometimes reply, "But everyone speaks English in India," by which of course they mean that the present market consists of people who speak English. If this is accepted, then to produce a localized version of a major operating system or office suite in Hindi would not only be extraordinarily expensive but useless, since "all computer users speak good English." The same is even more true for other South Asian languages, because each of them has fewer mother tongue speakers than Hindi and other Indian languages.

A related economic factor is the prevailing export orientation of the Indian software industry. To be sure, both the software and hardware associations of India have put localization at the top of their list of priorities. They insist that the great expansion of computer and Internet use to come in India will be domestic. If it is domestic, of course localization is required. But in fact, the orientation of the highly successful Indian software firms has been, so far, service-based, export-oriented, and therefore English-language based. One of India's greatest assets, reproduced in no other developing country, is its vast number of highly educated English-speaking computer designers and programmers. For this reason alone, nations like China, Russia, and Brazil, whatever their other strengths, will continue to find it difficult to compete with India in the software field.

These economic factors are powerful and in the short run decisive. But I am reminded of the story told by Harsh Kumar, the inventor of the localization system known as BharatBhasha. He tells of the two shoe salesmen who go to a remote Indian village with a population of one thousand people. The first salesman returns to his home office depressed and discouraged. "It is hopeless," he says, "there isn't a single person who wears shoes in the entire village." The second, however, returns jubilant and optimistic. "A wonderful opportunity," he says, "we can sell a thousand pairs of shoes."

Kumar insists that in the case of vernacular software, the absence of demand is created partly by the absence of supply. To take his favorite example, there are in Bombay hundreds, indeed thousands, of Marathi- and Gujerati-speaking merchants who own two or three shops and who currently spend every night until midnight balancing their books. They have the means and the need for computers that could do the job for them and get them home three hours earlier. But they do not have the command of English necessary to use any of the existing English-language small business packages. Computer consultants to whom they might turn can only offer English-language solutions, which are useless for the Marathi- or Gujerati-speaking merchant. The absence of supply automatically means the absence of demand.

At the very least, then, we need to examine critically the argument that economic factors alone suffice to explain the absence of local language software. Indeed there is a self-confirming quality to many economic arguments. If one asks why, in nineteenth-century Europe, there was no demand for video cassette recorders, the answer is simple: there were no video cassette recorders available. An analogous reply might go part way toward explaining the absence of demand for local language software: there can be no demand for a product which does not exist, or whose existence and utility is unknown. If local language software is not developed, or invisible, then the international software companies that claim that "there is no demand" will inevitably be correct.

A second factor that stands in the way of local language software is the very complexity—cultural, political, bureaucratic—of South Asia. One leader of a major American software firm, asked about localization to Indian languages said, "Okay, but which languages?" This is a reasonable question, but it has an answer: "Start with Hindi, go on to Bengali, Urdu, Tamil, Marathi, Telegu, etc." All of these languages are spoken by populations orders of magnitude larger than the populations of many nations for which locale coding or localization is currently available: for example, Norway, Denmark, or Latvia. Forward-looking companies, anticipating the steady growth of the vast Indian market, would be well advised to anticipate this market by localizing to major Indian languages. The winners in the next ten or twenty years in the Indian domestic market will be the firms that provide access to computers, Internet, and the Web in local languages.

Yet the complexity of the linguistic scene in South Asia points to the problem non-Indians (and some Indians as well) have in dealing with the subcontinent. India contrasts in this regard with the relative simplicity at the level of politics and written language of the other great Asian power, the People's Republic of China. In the latter, it is possible for American software firms to make binding agreements in Beijing for the use of the standardized written language that is employed by 1.3 billion Chinese. In India, for the many reasons suggested above, this is utterly impossible.

Other factors contribute to the slowness with which Indians and non-Indians alike have responded to the apparent potential of local language software. Among these is the fusion of language and power that has been at the center of this paper. The powerful in India, Pakistan, Sri Lanka, and Bangladesh are almost invariably those whose command of English is most perfect. Not only have they no personal incentive to encourage local language software, but, on the contrary, insofar as there is a class (or caste) interest in retaining power, it will be undermined by facilitating computer access to the non-English speaking, less powerful (and in India lower-caste) groups that already threaten the political hegemony of traditional Indian elites. I do not mean to suggest a conscious conspiracy, but only to propose that providing local language software to outcasts, tribals, scheduled castes, backward groups, slum-dwellers and other non-English-speaking local groups is unlikely to be paramount among the priorities of the powerful English speaking elites in South Asia.

Two other non-economic factors were once suggested by the head of a dynamic Indian software firm, who commented critically on a talk I once gave on local language computing. "You left out two of the central factors," he said, "the role of the Brahminical tradition and our ambivalent love affair with the English." By the first, he meant the traditional Brahmin emphasis on spirituality, transcendence, and higher orders of thought and action, contrasted with a distaste for all that is polluted, earthly, and material. "We are happy doing mathematics, astronomy, philosophy, and computers," he said, "but writing programs in Telegu or Hindi for the masses seems to many a less noble activity than programming in English or collaborating with a top-notch multinational firm in Germany." As for the "ambivalent love affair with the English," he referred to the embeddedness in modern Indian culture of formerly English games like cricket, the preservation amongst the Indian upper classes of clubs, schools, firms, institutions, and forms of government associated with the British, and above all, the continuing use of English as the prestige language of India. "It is one thing to program in English, which connects us to the wealthy, powerful and rich nations—to the rest of the world. But to program in Telegu, Tamil, or Marathi is to descend to the level of the street, to renounce the efforts of a century and a half to become English, to ally ourselves with the forces of primitivism in our nation" (Murthy 1997).

I cannot judge the validity of these arguments, but their claim is clearly that in addition to economic calculations, cultural factors play a role in the absence of vernacular software.

What Is To Be Done?

If local language software is important, and if it is largely absent in South Asia, the obvious question is, What is to be done?

Many wise men and women in India and elsewhere have answers to this question; mine will be a summary of theirs. First, however, I must note my disqualification: the solution to the problem of local software will obviously not come from American academics, but from the collaboration of South Asians in both public and private sectors interested in this problem, and perhaps from alliances with the multinational firms that today dominate the software market in South Asia. Here I can only offer a few suggestions.

The long-term potentials of the South Asian market need to be more accurately assessed. Although the present installed base of both telephones and computers is low in South Asia, the growth of the South Asian middle classes is rapid. Firms that project five, ten, or twenty years ahead are likely to be winners. Long-term projections could be the basis for rational economic investments in local language software.

In India, the role of the states will be central to localization. Existing policy in India requires the use of local languages in each state. As these states move toward the computerization of basic operations like electoral rolls, drivers' licenses, land records, or the interconnection by Internet of district offices, local language software will be necessary. This demand will probably precede and exceed the demand from individual computer owners. (In the United States, two-thirds of all PC sales are to institutions, not individuals.) Serving this market from the state governments will require major investments in local language software.

Standardization of language codes is a prerequisite for local language operating systems and applications. The Government of India, multinationals, and major Indian software firms need to cooperate in developing broadly accepted standards for the major Indian languages and in persuading programmers in India and abroad to use these same coding standards for each Indian language. ISCII may be adequate. But if, as some claim, ISCII has inadequacies, especially for the Southern Indian languages, then corrections need to be made rapidly. The standardization of local language codes needs to be a priority for the Government of India, and the several authorities of that Government that today deal with local language software need to be brought together and instructed to produce unified standards on a firm deadline.

Local language software and multimedia should be actively promoted both by the central Government of India and the governments of the local states. If local language "content" on the Internet and the Web continues to be absent, this will be an insuperable obstacle to local language information exchange. One positive role of government is to encourage (and finance, through start-up grants) projects that use local languages in education, in the development of databases, in Internet communication, and in multimedia Web-based projects. The current initiatives of the Government of Andhra Pradesh and Tamil Nadu stand as models of what other States and the Government of India might achieve.

Summary

The growing importance of digital technologies in South Asia reveals problems and opportunities for that region and lessons for other nations in the world. In South Asia are visible two issues critical for every nation on earth: how can the new electronic technologies be used to close, rather than widen, the gap between the powerful and the powerless, the privileged and the underprivileged? How can the new technologies be used to deepen, intensify and enrich the cultural diversity of the world rather than flatten or eliminate it? These questions come together with particular intensity in South Asia because of the fusion of power and language on that subcontinent. But by the same token, solutions that develop in South Asia will be relevant to the rest of the world. Just as India has been an example of how a developing nation can preserve democracy and cultural diversity, so South Asian solutions to the challenges of the Information Age could be a model for the rest of the world.

Notes

An earlier version of this paper was prepared for the Conference on Localization at the Center for Development of Advanced Computing, Poona, Maharashtra, India, in September, 1998. The research on which the paper is based is partially funded by a grant from the Nippon Electric Company, administered through the Provost's Fund at MIT. I am especially grateful to Patrick Hall of the Open University in England for his comments on this draft.

1. There is an extensive technical literature on localization. Typical are the works of Kano (1995) and Hall and Hudson (1997). A work that stresses cultural factors more than most is del Galdo and Nielson (1996).

2. See Rushdie and West (1999). For a contrary view that stresses the importance of fiction in Indian languages, see Mishra (1999).

3. Data on the precise numbers of speakers of Indian languages, or for that matter of any other language, are complicated by several factors.

One problem is the absence of agreement as to what is required for it to be said that an individual "speaks X language." How much fluency? How much ability to read and write? are required. Linguists offer no consistent answers to these questions. In a nation like India, where bi-, tri-, and quadrilingualism is common, the primary source for figures on Indian languages is *Ethnologue*. (See below.)

The second problem has to do with the inadequacy of studies of linguistic patterns and usage in South Asia. For example, the most comprehensive sources on linguistic patterns in South Asia are found in <http://www.sil.org/ ethnologue/countries/India.html>. But this document often relies on out-of-date figures (e.g., 1961 figures for English in India). Using more current figures, it indicates an extraordinarily low figure of 180 million primary mother-tongue speakers of Hindi (1991) and 346 million total Hindi speakers including second language users (1994). A recent article in the *New York Times* drawing on the *World Almanac* and 1990's figures puts Hindi speakers at 7.5X% of the world's population (of approximately six billion people), which works out to over 400 million. By this reckoning, Hindi closely follows English and is the third most commonly spoken language in the world. Other observers believe that Hindi speakers are more numerous than English speakers. All are agreed that Mandarin Chinese is far and away the most widely-used language (World Almanac 1999, 700f.).

Furthermore, with regard to languages like Mandarin or Hindi, no agreement exists on how to categorize dialects that may be mutually unintelligible variants of the "same" language or nominally different languages that are naturally intelligible. In India, some dialects of Hindi are said not to be mutually intelligible. And in South Asia, Hindi and Urdu derive from a common origin in spoken Hindustani. Urdu uses Persian script and has been deliberately "Persianized" by Muslims, and especially by Pakistani authorities, who have made Urdu a national language. (Before Partition virtually no one within the present boundaries of Pakistan spoke Urdu.) Hindi, in contrast, uses Devanagari script and has been to varying degrees "Sanskritized." Jawaharlal Nehru, whose native tongue was Hindi, complained that he could neither read the Indian Constitution in Hindi nor understand the Hindi broadcasts on Radio India because of the excessive Sanskritization of that language. See Wolpert (1996). The continuing congruity between Urdu and Hindi is shown by the enjoyment of Urdu television by Hindi speakers in northern India, and vice versa, and even more tellingly by the February, 1998 visit of Prime Minister Vajpajee of India to Pakistan. He addressed an Urdu-speaking Pakistani audience in Hindi, and, according to reports, was perfectly understood by the audience because of continuing similarities between Hindi and Urdu.

Similar imprecision exists with regard to the percentage of Indians who "speak English." The figure of 5% (approximately fifty million) is commonly accepted. But one commentator recently argued that only 2% "really" speak good English, while others have claimed that the percentage is as high as 10%. And for the purposes of computation, no one (to my knowledge) has studied how much proficiency in English is required in order to use a computer whose operating system, instructions, and interface are in English. Once again, some claim that one or two years of language training are adequate; other argue that in order to use any complex computer program, very high levels of English proficiency are needed. Finally, there is the question of English language e-mail and English language content on the Web.

Despite all these uncertainties, the overall linguistic pattern in South Asia is clear. In India alone, eighteen languages (including English and Sanskrit) are officially recognized. There are, according to the *Ethnologue* figures, thirty distinct languages in India with more than a million speakers. Certain linguistic groups like Hindi speakers are as large as the entire population of the European Union; Bengali, with an estimated two hundred million speakers, is approximately as common as French, Italian, and German combined. There are probably more Telegu speakers in Andhra Pradesh than there are German speakers in the world. The linguistic diversity of India, Pakistan, Bangladesh, Nepal, Sri Lanka, and the other South Asian nations is thus extraordinary.

But it is not unprecedented: among industrialized countries, Canada, Belgium, and Spain, to say nothing of the former Yugoslavia and the former Soviet Union, have very large linguistic subcommunities. The great majority of sub-Saharan African nations like Nigeria, Kenya, or South Africa, have multiple linguistic communities. Indeed, the monolinguistic pattern of the United States, where more than 95% of all inhabitants speak good English, is highly exceptional and perhaps even unique on the world scale.

4. The linguistic history of South Asia is complex and largely unanalyzed. Early works by Fishman et al. (1968), Das Gupta (1970), and by Brass (1974), lay out general issues as of twenty-five years ago. Laitin (1992) focuses on Africa, but uses the Indian model of a colonial language, a national language, plus a local language as the paradigm for Africa as well. Laitin assumes that the colonial language (e.g., English or French) is part of the national linguistic repertoire, but in the case of India and presumably most African nations, this is true only of a small cosmopolitan elite.

More recent works include King (1997) and Tariq Rahman's excellent work (1996), which chronicles at length the role of language in the East/West Pakistan war that led to the creation of Bangladesh. On the militancy with regard to the use of the Tamil language, see Ramaswamy (1997).

5. On the history of British English language policy in India, see Read and Fisher (1998), and Viswanathan (1989).

6. Communication theorists discuss this as the Sapir-Whorf Hypothesis. See, for example, Gudykunst and Kim (1997), and Griffin (1994). I am indebted to Charles Ess for these references.

7. See <http://www.elcot.com/tamilnet99.htm> (International Seminar on the Use of Tamil in IT, Chennai, February 7–8, 1999).

8. Schon et al (1999) and Eisner (1999). Eisner notes, "Seventy-five percent of households with incomes over $75,000 own computers, yet only 10% of the poorest families in this country [the United States] have computers." Eisner cites data from *USA Today* (no date).

References

Barber, Benjamin. 1992. "Jihad versus McWorld." *Atlantic Monthly* (March), 53–63.

———. 1995. *Jihad versus McWorld*. New York: Times Books.

Brake, David. 1996. "The U.S. Wide Web." *Salon* (Apple on-line), Issue 30 (September 3–6).

Brass, Paul. 1974. *Language, Religion, and Politics in North India*. New York: Cambridge University Press.

Das Gupta, Jyotirindra. 1970. *Language Conflict and National Development: Group Politics and National Policy in India*. Berkeley: University of California Press.

del Galdo, Elisa, and Jakob Nielson. 1996. *International User Interfaces*. New York: Wiley.

Eisner, David. 1999. "The Social Impact of the Internet: A Commercial Perspective." Draft presented at June 4 meeting of the National Academy of Science/Max Planck Institute Task Force on "Global Networks and Local Values."

Fishman, Joshua, Charles Ferguson, and Jyotirindra Das Gupta. 1968. *Language Problems of Developing Nations*. New York: Wiley.

Griffin, E. M. 1994. *A First Look at Communication Theory*. New York: McGraw Hill.

Gudykunst, William B., and Young Yun Kim. 1997. *Communicating with Strangers: An Approach to Intercultural Communication* 3rd ed. New York: McGraw Hill.

Hall, P. A.V., and R. Hudson. 1997. *Software Without Frontiers*. Chichester, England: John Wiley and Sons.

Kano, Nadine. 1995. "Developing International Software for Windows 95 and Windows NT." Redmond, WA: Microsoft Press.

Keniston, Kenneth. 1997. *Software Localization: Notes on Technology and Culture* (Working Paper #26). Boston: Program in Science, Technology, and Society, Massachusetts Institute of Technology.

King, Robert D. 1997. *Nehru and the Language Politics of India*. Delhi: Oxford University Press.

Laitin, David. 1992. *Language Repertoires and State Construction in Africa*. New York: Cambridge University Press.

Mishra, Pankaj. 1999. "A Spirit of Their Own." *New York Review of Books* (May 20), 47–53.

Murthy, Nayaran (CEO of Infosys). 1997. Remarks made at the National Institute of Advanced Studies, Bangalore, India, March.

Rahman, Tariq. 1996. *Language and Politics in Pakistan*. Karachi: Oxford University Press.

Ramaswamy, Sumathi. 1997. *Passions of the Tongue. Language in Tamil India, 1891–1970*. Berkeley: University of California Press.

Read, Anthony, and David Fisher. 1998. *The Proudest Day: India's Long Road to Independence*. New York: Norton.

Rushdie, Salman, and Elizabeth West, eds. 1999. *Mirror Work: 50 Years of Indian Writing, 1947–1997*. New York: Henry Holt.

Schon, Donald, Bish Sanyal, and William J. Mitchell, eds. 1999. *High Technology in Low Income Communities: Prospects for the Positive Use of Advanced Information Technology.* Cambridge, MA: MIT Press.

Viswanathan, Gauri. 1989. *Masks of Conquest: Literary Study and British Rule in India*. New York: Columbia University Press.

Wolpert, Stanley. 1996. *Nehru: A Tryst with Destiny*. New York: Oxford University Press.

The World Almanac and Book of Facts, 1999. Mahwah, NJ: World Almanac Books.

Global Culture, Local Cultures, and the Internet: The Thai Example

Soraj Hongladarom

Introduction

The growth of the Internet is a world-wide phenomenon. From a relatively obscure academic tool, the Internet has become a household fixture and now it is hard to find anyone without an e-mail address or a personal home page. Cyberatlas (<http://www.cyberatlas.com/geographics.html>) reports by pinging 1% of all the Internet hosts that in January 1996 there were 9,472,000 distinct hosts, and 16,146,000 in January 1997, an increase of 170%. As more and more people are becoming wired, the Internet itself is fast becoming as pervasive as televisions and radios. However, its ability to generate many-to-many communication sets it apart from these traditional mass media. This gives the Internet a strong potential in forming communities, and where there are communities, there are cultures unique to each community. The potential of the Internet in forming "virtual" communities incurs a number of problems, chief among which is the relation between the community formed by the Internet itself and the existing communities bound by locality and cultural tradition.

The Internet at the moment is still predominantly American, but it is increasingly global, with more and more countries adding more and more host machines, expanding the network at a breathtaking speed. Network Wizard (<http://www.nw.com/>) reports that the growth of Internet hosts in 1994 was 15% in Asia alone, and in Thailand the growth rate was as much as 53%. This expansion has created a problem of how local cultures adapt themselves to this novelty. As a quintessentially Western product, there is clearly bound to be a contrast, if not necessarily a conflict, between non-Western cultures and the Internet technologies. How, in particular, do local cultures take to the Internet and other forms of computer-mediated communication such as the Bulletin Board System (BBS)? Does the Internet represent an all unifying force, turning all cultures within its domain into one giant superculture where everything becomes the same? Does the idea of the Internet and other forms of computer-mediated communication carry with it cultural baggage of the West, such as democracy and individualism?

This paper attempts to provide some tentative answers to these vexing questions. It presents a case study of one local culture, that of Thailand, in computer-mediated communication. More specifically, it presents a case study of the Usenet newsgroup on Thailand and its culture, soc.culture.thai, in order to find out whether and, if so, how Thai cultural presuppositions affect the received underlying ideas of the CMC technologies. Then we shall see how these answers provide an insight into the theoretical problem of the extent to which global computer-mediated communication could be regarded as a means to realization of such Western ideals as liberalism, individualism, respect for human rights, and democracy.

I argue in this paper that Thai cultural attitudes do affect computer-mediated communication in a meaningful way. This means the idea that the Internet would automatically bring about social change in line with developments in the West needs to be critically examined. It appears from the study that important presuppositions of local cultures are very much alive, and exist alongside the imported Western ideas. Which type of cultural attitudes and presuppositions is present is more a matter of pragmatic concern, such as whether the participants in CMC happen to find any use for a set of ideas, than that of truth or falsity of the ideas in questions.

Internet in Thailand

Kanchit Malaivongs reports (<http://203.148.255.222/cpi/it4.htm>) that Internet connection in Thailand first took shape in 1988 when an e-mail-only dial-up account was set up between Prince of Songkhla University in southern Thailand and the Australian Academic and Research Network (AARNET) through the help of the Australian government. A few years later in July 1992, Chulalongkorn University set up the first permanent leased line connection and provided services to faculty and students of the university as well as those of some other participating universities. The cost of connection was shared among the universities, and faculties and students enjoyed free access. Another permanent connection to the Internet backbone was set up by the National Electronics and Computer Technology Center (NECTEC), a government agency responsible for computer and information technology issues, and more academic organizations joined. Soon the government decided to open up access to the general public and dozens of commercial Internet Service Providers (ISPs) sprang up. Today it is estimated that around 131,000 Thais are enjoying access to the Internet (*Phuu Jad Kaan Raai Wan* 1997, 30).

soc.culture.thai - Wild Frontier of Things Thai

For the majority of Thai Net surfers, soc.culture.thai (SCT) is by far the most popular Usenet newsgroup. It is perhaps the place in cyberspace for discussion on all sorts of aspects on Thailand, and it deals with all aspects of Thai society and culture. Thais form the majority of the nationalities of discussants in the group.[1] The newsgroup derives its tremendous popularity among Thais and Thai watchers from its free-wheeling threads of discussion in a culture where some topics may not be discussed publicly. Furthermore, the group also serves as a place where struggles for political freedom take place, a phenomenon also reported by Andreas Harsono (1997) in the case of Indonesia. During the May 1992 incident, when soldiers opened fire on the Thai people fighting for constitutional reform, the newsgroup was one of the means of struggle. The whole world was kept informed of what actually happened, and many Thais who were locked out of reliable information due to government blackout of the national media relied on it to learn what was happening outside their homes. Now that the political climate is much freer, the newsgroup still remains politically active. Members of the newsgroup cherish the freedom to openly discuss forbidden topics with fellow members. Such topics include the personal character of the members of the royal family, and criticisms—or in many cases, invectives—against the politicians.

Since Thais can apparently talk and discuss freely on the Internet without fear of reprisal from the authorities, it is understandable that they would want the same amount of freedom outside of the newsgroup, too. What is emerging from many discussion threads in the group is that the participants want to see a new Thailand which is more open and more in tune with the world community—

a country that is less bound to the past while still retaining its own cultural identity. An example can be seen from a particular thread, "The king said new constitution is acceptable." The thread started from an important event in recent Thai history, when the King signed the new constitution into law. Discussion then ensued in SCT concerning the new constitution. Naturally the discussants hoped that the new constitution would bring a new era in Thai politics, an era when the old dirty, vote-buying, voters-bullying, raw power politics would be over. There were some disagreements, however. One rather controversial point in the new constitution concerns the qualifications of those who are to enter politics. Candidates for parliamentary election are now required to possess a minimum of a bachelor's degree. The rationale of the Constitution Drafting Assembly, the organization responsible for drafting the new charter, was clearly to react against the current situation where many powerful MPs and hence cabinet members do not have the necessary knowledge and skills for running the country. As a result, these leaders often act as if they represent their constituencies only and do not have a broader view of the country as a whole.

However, a significant number of SCT members voiced their disagreement with the clause. A member, Prapasri Rajatapiti, writes:

> That the one issue I have been strongly opposing for the new constitution. I for one believe these articles to be very discriminatory. I believe that as long as one can read and write, one can serve as an MP. Education is only compulsary up to grade 6. How can we tell these people who did not have the chance to go to school, and was told that it was OK then (since it is not compulsary), that now they won't have a chance to be MP or senator unless they go back to school. Formal education is only 1 form of education, not all.[2]

As usual for threads of discussion, Prapasri's argument did not go unopposed. Another contributor, giving only his personal name, Tirachart, raised exactly the same point as the CDA on the ability of undereducated politicians to run the country:

> Hello;
>
> It's about time to change or else Mr. Cow and Mr. Kwai will be minister of something. Does it make you happy to see the

> government's way of serve the people nowaday? How much longer those jerk will be still incharge the of goverment?[3]

"Cow" is English, and "Kwai" is a Thai word meaning "water buffalo." In Thai language, to call someone a cow or a buffalo means that he or she is stupid. This kind of venting of emotion is common in SCT. Here one can find that flaming the government and politicians is among the most favorite pastimes. The more virulent the attack, the higher "status" the attacker seems to possess in the group. Tirachart's post here is also interesting in that it presupposes some cultural background in order to understand it fully. Without the knowledge that Thais perceive bovines to be very stupid, non-Thais have to rely on contexts to guess the meaning, but sometimes this is quite difficult.

In fact, comparing the politicians with animals is rather common. Commenting on an earlier post by Sanpawat Kantabutra, one calling himself "Aitui" writes:

> On 17 Oct 1997 01:18:35 GMT, sanpawat@c4.cs.tufts.edu (Sanpawat Kantabutra) wrote:
>
> >I believe so. It will take about 25-30 years for younger generation
>
> >like us will be in major positions in the government and other state
>
> >organizations. I think the new generation is better than the old one
>
> >in terms of ... Well, almost everything. Khun Anand also said that it
>
> >is the time for younger generations to run Thailand. 25-30 years are
>
> >worth-while.
>
> >We dun need 25-30 years...just kill those fucking heas then we will have a much better tomorrow ![4]

This is more of an expression of anger than a deliberation. However, the rationale behind it is clear. Sanpawat comments that the next generations of Thais would be more qualified and more responsible than the present one, presumably due to better education and more openness. "Hea" means "monitor lizard," a much lower-ranking animal in the Thai cosmos than bovines. While bovines are merely stupid, monitor lizards are treacherous and evil. Bovines are viewed by Thais as beneficial, as they help them with tilling the fields. Many Thais feel a certain sense of gratitude to them. Monitor lizards, on the other hand, are always keen to steal the farmers' chickens and ducks. The word "hea" is in Thai a strong invective used to describe those who are bad and depraved.

By mixing Thai words in the more or less English posts in SCT, the contributors do not as much aim at being fully understood by the global community than at talking and sharing feelings within their rather close-knit community. Here those who do not happen to understand these words and the presupposed background knowledge necessary for grasping the whole meaning feel left out. Thus, SCT takes on a double function. On the one hand, it acts as a channel of disseminating information about Thailand and its people, as stated in its charter. On the other, it serves as a means by which Thai people and non-Thais who are "in the know" strengthen their shared feelings and knowledge. It is as if the newsgroup is a coffeehouse where people who know one another very well come to discuss things in which they are interested. They do not quite care whether outsiders would be able to follow what is going on. That is not the point of the communication. Such a communication as happening here has its essential function within a community. SCT, in this instance, is the place where members of the community come to share views, thoughts and feelings, thus making the community itself possible.

This view of communication as the means of strengthening community ties is called by James Carey the "ritual" view. In *Communication as Culture* (1989, 18–23), Carey states that there are two views on communication, namely the "transmission" and the "ritual" views. The former views communication as one-way traffic, where information, injunctions, news, and the like are "transmitted" from the source of power to remote posts. One purpose of such transmission is to create political unity and to assert the power of the political center to areas within its jurisdiction. The ritual view, on the other hand, views communication not primarily as a means of transmitting information, but as an integral part of community activity, which members of a community perform in order to reaffirm the identity of the community itself.

The invectives against the Thai political leaders in the SCT are parts of the government bashing occurring after the great flowering of media freedom following the Black May Incident of 1992.[5]

Released from the fear of criticizing the authorities, Thais began to view the government not as something from far above, but as an institution of their own. Once they feel that criticizing the government incurs no real threat to their safety and liberty, Thais enjoy this freedom a lot, and sometimes it may seem that the criticisms serve merely to release pent-up emotions and frustrations rather than to offer constructive viewpoints toward solving the country's problems. What is rather surprising in this phenomenon is that not only highly-educated, middle-class Thais are joining in this bashing frenzy, but the poor farmers in the countryside are joining the fray, too. Traditionally these poor farmers, who form the majority of the Thai population, have a very high respect and awe for their rulers, including political leaders and bureaucrats. But they are beginning to feel, in the more democratic and liberal climate, that the leaders are merely humans, and most importantly that they themselves do have real power and leverage against them. Since these leaders do come to power only through their votes, the villagers are getting more involved in politics; they are trying to wrench back power to take care of their own affairs from the bureaucrats. A new community is emerging that is bound by a sense of independence and increasing responsibility in dealing with one's own affairs.[6]

Another thread in the newsgroup from which we can see cultural implications concerns the use of language in postings. Kritchai Quanchairat, a regular contributor to the newsgroup, is a Thai computer scientist specializing in localizing certain Internet softwares. He is known for his campaign for more postings in Thai language. Naturally his campaign provoked a fair number of replies. In a post replying to Kritchai's, "Conrad" writes:

In article <199709122354.SAA27681@phil.digitaladvantage.net>, kritchai

Quanchairut <kritchai@usa.net> writes

>[You may use Thai or English as you prefer on SCT/TMG]

>

>I linked posts from TMG to Soc.culture.thai.

>I beleive posts in Thai will help most of soc.culture.thai

>readers (who are the majority behind the scence in Thailand)

>to be able to ACTUALLY MAKE USE OF THE INTERNET.

>

>Most Thais could not read English very well if not at all.

>These will most benefit those K12 kids who are getting on-line

>via SchoolNet projects. It's not too late to help the kids

>to get on-line today. Some of us may need to be a little

>patience about this. Let's think of it as "FOR THE KIDS".

>

>If you don't know how frustrating it is for not being able

>to read/understand posts in their own groups, check German

>or French groups.

>

>It's time and your open-mindedness counts!!

>

>Krit

>...

>

I was under the impression that this n/g was created to discuss and disseminate aspects of the Thai culture, social and political scene. The vast majority of people using this n/g do not read/write Thai so posts in Thai will restrict the original purpose. By all means set up a Thai language n/g. It is a fact that the common language of the internet is English, being either the first or second language of the majority of users. Surely it is a desirable aim that the information on the internet should be accessible to the widest possible audience.

To progress academically, socially and economically in Thailand one MUST be competent in the English language. What better incentive could there be for kids who wish to join the on-line community?[7]

Kritchai's attempt to persuade SCT members to use Thai in their posts amounts to nothing less than changing the whole face of the group. However, he has a point. The level of English understanding in the country is generally poor, and the language is not in widespread use at all. Proficient users of the language are few compared to the whole population. Thus, Kritchai apparently believes that if Thai is used more in SCT, more Thais would be persuaded to join and the ensuing discussions would be good for them.

Another reason in favor of using Thai in SCT concerns power relation among different language speakers, as implicitly stated in Kritchai's post. Thais sometimes feel it unfair that they have to communicate in a foreign language instead of their own; they often feel inferior to native English speakers just because their English is not as efficient at enabling them to talk as fast or to argue as effectively as the natives. Using Thai in this context amounts to an empowering of non-English speakers so they feel confident enough and less self-conscious enough to participate actively in the newsgroup. Since English has never gained a foothold in the country except as a foreign language, many Thais feel resistant to the idea of having to talk in English on matters of themselves and their culture. They do not feel that SCT is a forum about Thailand and its culture, but they appear to feel that it is also for Thais and sometimes Thais only. In a tight, close-knit culture such as the Thai one, such feelings are not uncommon.

Internet as Globalizing Agent?

Let us return to our original questions. Does the Internet succeed in turning all cultures of the world into one monolithic culture where all the important beliefs and background assumptions are the same? In one sense, it would appear so. When participants of widely disparate cultures come to interact, what happens is that there emerges a kind of culture which is devoid of historical backgrounds that give each local culture its separate identity; it is, for example, the culture of international conferences. The newly-emerging culture is comparable to piped music one hears in airports or in modern supermarkets; that is, it is shorn of its value, its role in a people's scheme of things. It plays no part in the ritual of a traditional culture. In short, it has become sanitized and modernized. Let us call this kind of culture the "cosmopolitan" one. One aspect of the Internet clearly points to that kind of culture. When people from all parts of the globe communicate with one another, it is difficult enough when they face each other to observe all the non-verbal cues. (Those cues might be interpreted differently.) But since Internet communication happens almost exclusively through texts, the task becomes much more difficult. Communication requires that participants share at least some sets of values and assumptions. Participants have to accept that what others say are largely true, as Donald Davidson (1984, 200–01) argues. Thus when texts become the only means of communication in building a virtual community, this shared set of assumptions and values already exists. These values, however, do not belong to any local, traditional culture, but are whatever makes global computer-mediated communication possible.

It is well known that the shared set of values and assumptions prevalent on the Internet resembles that of liberalism and egalitarianism typical of modern Western, liberal culture. The origin of the Internet as a repository for exchanges of discussion and information by computer scientists and other scholars points to the fact that the Internet bears the stamp of the culture of this group. Its birthplace in the United States explains why these assumptions and values are so well-embedded. Nonetheless, the potential of the Internet as the global forum of international communication makes it almost necessary that this shared set of values and assumptions is held by the participants. The set is an outcome of an international, cosmopolitan culture where participants share little in common in terms of historical backgrounds. In order to make communication possible among those who come from disparate historical, traditional backgrounds, the values and assumptions germane to a particular local culture cannot do the job. Participants either talk about their professional matters, the topic of international conferences, or they talk about superficial stuff guaranteed to be shared already, like the weather. The Internet does not have to originate in the United States for it to acquire the cultural traits it already has. It could have come from Japan, for example, but when it is truly globalized it has no choice but to be what it is now. It is in this sense, then, that the Internet could be regarded as a globalizing agent.

This shared set of values and assumptions typical of the Internet becomes apparent when it spreads its roots to states where the ideas of liberalism, egalitarianism, and democracy face violent resistance from the political authorities. The newsgroup soc.culture.burma, for example, is used by Burmese dissidents living abroad to spread information which would not be known otherwise. It is no surprise that the Burmese government even requires its citizens to ask for official permission to own a modem. Failure to do so can make one a political prisoner. That is what happens when governments actively attempt to stop the wishes of its people, and it shows how potent the Internet can be as a political force.[8] It also shows that, if we take the ideas of democracy and respect for human rights as universal, then the Internet could be seen as a harbinger of these ideals to the areas where the ideals are not appreciated by the authorities.

This aspect of the Internet as a harbinger of the liberal ideals could be taken to substantiate the claim that the Internet represents a global force spreading Western values to the world, as if it were the destiny of the world to subscribe fully to Western ideals. However, I think a distinction should be made between Western culture and cosmopolitan culture. Western culture is a product of more than two thousand years of continuously-evolving civilization. It has its own traditions, customs, belief systems, and religions, putting it on a par with the world's other great civilizations, such as India or China. Cosmopolitan culture, on the other hand, is borne out of the need for people from different cultures to communicate or to do other things with one another. Thus it is by nature shorn of any resources that could be drawn from centuries of experiences. What is happening with the Internet is perhaps not a spread of the former, but the latter. But that is hardly surprising. It is true that cosmopolitan culture originated first in the West, because the need for finding common ground among people of disparate beliefs was first felt there; that, however, does not mean that the two cultures are one and the same.

Thus, when the Internet is used as a political tool, it does not necessarily mean that it acts as a Westernizing force. The majority of SCT contributors who criticize the Thai government are Thais, and here the newsgroup could have been a traditional Thai coffeehouse where people gather and talk and discuss politics. The participants in the newsgroup do not become less Thai when they surf in cyberspace. Instead as they become more active in the affairs of their country, and they show that they are more attached to their locales. Moreover, as the Thai participants can use, and have indeed used, the Internet to spread information on various aspects of their culture, such as traditional recipes and digitized traditional music and paintings, the Internet can even be a tool for cultural preservation and propagation. In this sense it does not globalize, but localize, making people more attuned to their own cultural heritage. Nonetheless, as an embodiment of cosmopolitan culture, it is clear that the Internet globalizes in this way—as a means by which global communication and community-building, if only "thinly" in Michael Walzer's (1994) sense, becomes possible.

According to Walzer, moral arguments are "thin" when they lose the particular histories and other cultural embodiments that make them integral parts of a cultural entity. These are the parts that make the arguments "thick." To use Walzer's own example, when Americans watched Czechs carry placards bearing words like "Truth" and "Justice," they could relate immediately to the situation and sympathized with the marchers. However, when the arguments are at the local level, as to which version of distributive justice should be in place, there might well be disagreements, and Americans may find themselves disagreeing with the particular conception of justice which is eventually adopted. The sympathetic feeling one feels across the Ocean is part of the "thin" morality, but the localized and contextualized working of those moral concepts is part of the "thick" (Walzer 1994, 1–19).

The thread of discussion in SCT concerning the language to be used in the forum illustrates the tension between local and global cultures, or thick and thin conceptions, very well. As usually happens in international conferences, talking only about the weather to those with whom one does not share much is rather boring. Many non-Thai Internet surfers do not know much about Thailand and the variously subtle nuances of her culture; their contributions therefore are generally limited to asking for information, and when they venture to provide information or ideas of their own, they often reveal that they are quite ignorant of the deeply-rooted culture. In order to communicate with non-Thais on topics related to Thai culture, Thais have to supply an adequate amount of background information in order to make themselves understood. It is much easier for them just to talk to fellow Thais who already share such background knowledge. This way they can mix Thai words in the posts, refer to "kwais" or "heas," or allude to characters in the classical literature without fear of not being understood. Consequently, participants in international gatherings sometimes drift off to form their own smaller groups, banding with those to whom they share background knowledge. The situation also happens on the Internet. The founding charter of SCT states that the newsgroup is created in order to exchange information and viewpoints about Thailand and its culture, and that English is to be the only medium of communication.[9] But since most Thais do not use English very well, the campaign to post in Thai language is understandable. There also has been an attempt to amend the SCT charter to make it officially acceptable to post in Thai. The implication this debate has for the issue of the Internet as a globalizing force is clear.

The ongoing debate in SCT on what language is to be used, together with the de facto existence of a significant portion of SCT posts which are entirely in the Thai language, provide evidence that, instead of looking at the Internet as a sign of the world becoming culturally monolithic, we may have to look at it just as a global forum where participants join one another so long as there is a felt need for it. And when they feel more comfortable talking to someone back home, so to speak, they feel no qualms in forming smaller groups within the larger gathering, where they can forget the learned *lingua franca* and enjoy talking in the vernacular. To assume that the Internet would bring about a culturally monolithic world would mean that it would bring about a set of shared assumptions and values, including respect for human rights, individualism, egalitarianism—in other words, the ideas of contemporary liberal democratic culture. But since it is conceivable that those liberal ideals could exist within cultures other than those of the West, to claim that the Internet would bring about the same "thick" culture in Walzer's sense would seem to be mistaken. If the set of ideals is viewed instead as a part of the cosmopolitan culture, then it appears that the set will be adopted by a local culture if it feels that it wants or needs to be a part of the global community. And if members do not feel the need, then they will just turn their back on it, in effect telling the world that they do not care to join. Very often in those cases, the wish of the populace runs counter to that of the political leaders; political oppression and prohibition of freedom of expression result.

If the culture believed to be "exported" by the Internet is viewed as a cosmopolitan one, and not the traditional Western culture, then we are in a good position to assess the claim that the Internet is a homogenizing cultural force. Since cosmopolitan culture is neutral on most respects, the claim that the Internet will bring it about is rather trivial. On the other hand, if traditional—or Walzer's "thick"—culture is at issue, then it seems the Internet fails to provide such a culture. But now the crucial question is: to which culture do the salient aspects of modern liberal culture, namely respect for human rights, democracy, egalitarianism, belong? Do they belong to the traditional Western culture, putting them on a par with Christianity, the Gothic cathedrals, Bach's chorales, Michelangelo's paintings, Franz Kafka's stories—in short with the aspects that give Western civilization its uniqueness? Or do they belong to secular, cosmopolitan culture, the culture arising out of the need of people from various cultures to get in touch with one another? To answer this question deeply enough and satisfactorily enough would itself require at least another paper. But at least a glimpse of the way toward an answer can be given here. We have seen from the examination above of what happens in SCT that it is certainly possible for Thais to fight for democracy and human rights while retaining their distinct cultural identities. The invectives against the government are just some indications of the concerns of the Thai people for their government and their own country; behind an invective lies a vision of how the country should be governed, a vision that does not include the current political leaders. On the other hand, the debate on the language to be used in the newsgroup shows that Thais are conscious of their identities and the need to form their own smaller communities within the globalized cyberspace. That the threads happen together in the same newsgroup shows that Thais do not view the struggle for more openness, more efficient government, more participatory democracy and so on as something separated or incompatible from the desire to assert their cultural identity. There is no necessary conflict between these two spheres of culture, in the same way as there is no necessary conflict between Bach's chorales and the Gothic buildings on the one hand, and the democratic, libertarian, and egalitarian ideals on the other.

Conclusion

Thai attitudes toward CMC technologies, especially the Internet, seem to show that the technologies only serve as a means to make communication possible, communication that would take place anyway in some other form if not on the Internet. Most Thais welcome the new technologies, thinking that they enable them to surge forward with the world. However, this is a far cry from claiming that the Internet brings about a culturally monolithic world where everybody shares the same "thick" backgrounds and values. In the SCT newsgroup, Thai people and non-Thais talk about matters that are interesting to them, be they politics, culture, etc. Here the newsgroups act more like the traditional Thai coffeehouse where public matters, especially local and national politics, dominate the discussion. As the Internet is really a form of the media, and in Thailand it has been heavily promoted that way, it is an open to the world at large, where, to paraphrase Marshall McLuhan, one can extend one's senses far from what is normally possible. One can perceive what is going on in far corners of the world in an instant, and one can feel as though one is bodily transported to the remote regions with whom one is interacting.

What comes naturally from such a scenario is that there are bound to be comparisons between what one perceives in the far corners and in the local areas around oneself. When one sees in the far corners what one believes to be good for one's own locality, it is natural to suppose that there are going to be changes in the latter. Richard Rorty argues that the process is what actually lies behind the universalist rhetoric claiming for a common morality and social norms for all mankind. This process of changes in one's locality as a result of one's perception of other regions, according to Rorty, should not be taken to imply that there is a universal ethics at work. Rorty's naturalism would make such ethics redundant. What is really the case is that some people just want to live like others. Thus, instead of a universal consciousness that this is the right way to live, Rorty claims that there is "solidarity" for mankind (1989, 1991). Hence, when a Thai Internet surfer sees what is going on in another region of the globe which she believes would be good for her own country, be it the strict enforcement of the law, open democracy, human rights, or so on, she wants to be a part of the community that she finds acceptable. Since she can decide freely on her own, there is no need for her to change her own cultural identity. She can remain Thai while embracing all these political and social ideals. That is to say, she can enjoy Thai food and Thai music while struggling for a more open democracy in Thailand at the same time.

Thus the Internet and local cultures both determine each other. While the Internet is a window to the world where influences can be received, the content of the Internet is obviously determined by what is posted or uploaded to interconnected computers. The information available shows that cultural groups are as separated from one another as they are in the outside world. The cultural fault lines, so to speak, stay roughly the same. An outsider would feel as much lost in the cyberspace of SCT as they would be when dropped in the midst of a Thai town. According to Carey's ritual view, communication is one of the rituals of a culture that give it its uniqueness, its being. Hence communication in SCT could be seen as one of the rituals that make up the Thai identity. The identity, however, is not something static, but is constantly evolving so as to respond effectively to outside changes. Thus there is no contradiction in saying that the Thai identity, for example, evolves in such a way that the Thai people accept ideals such as human rights, democracy, and the like as their own, as integral parts of their culture. Cyberspace mirrors real space, and vice versa.

Notes

Travel grant for the London Conference was supported in part by the Faculty of Arts, Chulalongkorn University. The author wishes to thank Dr. M. R. Kalaya Tingsabadh, Dean of the Faculty, and Assoc. Prof. Thanomnuan O-charoen, Deputy Dean for Academic and Research Affairs, for their generous help and support.

This chapter appeared originally in the *Electronic Journal of Communication / La revue electronique de communication*, 8 (3 & 4), 1998 (see <http://www.cios.org/www/ejcrec2.htm>), and in *AI and Society* (1999) 13: 389–401, and is reprinted by kind permission of the editors and publishers.

1. According to the soc.culture.thai General FAQ (available online at <ftp://rtfm.mit.edu/pub/usenet/ soc.culture.thai>), a survey in 1994 shows that soc.culture.thai has an estimated readers of 39,000 worldwide; 66% of all USENET sites carry this newsgroup; and total monthly traffic is 2035 messages or 4.4 MB. Thais form the majority of those who read and post in the newsgroup, comprising 64% of the total.

2. Prapasri Rajatapiti, post to soc.culture.thai, message-ID: <19971010230101.TAA17707@ladder02.news.aol.com>, October 10, 1997.

3. Tirachart, post to soc.culture.thai, message-ID: <61mk27$6s5$1@excalibur.flash.net>, October 10, 1997.

4. Aitui, post to soc.culture.thai, message-ID: <34474ee0.8970100@news>, October 17, 1997.

5. Anek Laothammatas (1993) argues that the urban middle class were the key players in the demonstration, making it different from the previous ones which had been led by student activists. He points out that the middle class would like to see a transparent government which is free from corrupt practices and a more modern, more open political system. This wish of the middle class is clearly reflected in the tones of most discussions on Thailand on the Internet.

6. However, since the middle class have the economic and cultural power, their voices in the affairs of the country is very loud indeed, and cannot be fairly compared to that of the villagers. Moreover, since the number

of Thai people connected to the Internet are currently very limited, and the fees for a connection is far from affordable, members of the Thai Internet community consist solely of the middle class. For them the Internet has become an important tool by which they create and maintain a community. One aspect of this community is that the members agree that old style politics needs to change, and that Thailand needs to open herself up more and become more an open, liberalized society.

7. Conrad, post to soc.culture.thai, message ID: <3pH6RKAMWmG0Ew8t@ceebees.demon.co.uk>, September 13, 1997.

8. The relation between Internet and democracy appears to be parochial. It depends on the situations where a particular communication/ community takes place. For Thailand, the fight is for more open, more transparent and efficient government. For the US, the situation might be as described in Mark Poster in "Cyberdemocracy: Internet and the Public Sphere" (1997, 201–217). That is, Poster calls for a kind of 'postmodern' or more participatory democracy, which is less encumbered by the traditional forms of American government. This seems to show that the Internet is more a tool for those who need it than a homogenizing force, making every culture the same.

9. Soc.culture.thai general FAQ, available online at <ftp://rtfm.mit.edu/ pub/usenet/soc.culture.thai>.

References

Anek L. 1993. *Mob Mue Thue: Chon Chan Klaang Lae Nak Thurakij Kap Pattanaakaan Prachaathippatai* ("Mobile Phone Mob: The Middle Class, Businessmen, and the Development of Democracy"). Bangkok: Matichon Press.

Carey, J. W. 1989. *Communication as Culture: Essays on Media and Society*. Boston: Unwin Hyman.

Davidson, D. 1984. *Inquiries into Truth and Interpretation*. Oxford: Clarendon.

Harsono, A. 1997. "Indonesia: From Mainstream to Alternative Media." *First Monday: A Peer-Review Journal on the Internet*, available at <http://www.firstmonday.dk/issues/issue3/harsono/>.

Malaivongs, Kanchit. 1997. "Internet Services in Thailand," available at <http://203.148.255.222/cpi/it4.htm>.

Phuu Jad Kaan Raai Wan (Manager's Daily). 1997. 6 October, 30.

Poster, M. 1997. "Cyberdemocracy: Internet and the Public Sphere." In *Internet Culture*, ed. D. Porter, 201–17. New York: Routledge.

Rorty, R. 1989. "Solidarity." In *Contingency, Irony, and Solidarity*, 189–98. Cambridge: Cambridge University Press.

Rorty, R. 1991. "Solidarity or Objectivity?" In *Objectivity, Relativism, and Truth: Philosophical Papers*, 21–45. Cambridge: Cambridge University Press.

Walzer, M. 1994. *Thick and Thin: Moral Argument at Home and Abroad*. Notre Dame, IN: University of Notre Dame Press.

Contributors

Charles Ess is Professor of Philosophy and Religion and Director of the Center for Interdisciplinary Studies, Drury University (Springfield, Missouri). He has published in ethics, history of philosophy, feminist biblical studies, contemporary Continental philosophy, computer resources for humanists, and computer-mediated communication. His book *Philosophical Perspectives on Computer-Mediated Communication* (SUNY Press, 1996) is used nationally and abroad as a textbook. Ess has received awards for teaching excellence, as well as a national award for his work in hypermedia.

While a Research Associate at the Center for the Advancement of Applied Ethics, Carnegie Mellon University, Ess organized on-line dialogues and contributed to the CD-ROM, *The Issue of Abortion in America* (Routledge, 1998). With colleagues at Drury in architecture and philosophy, he is exploring ways of enriching both disciplines with the pedagogies and insights of each. He has traveled in the Middle East and lived in several European countries. He and his wife Conni, an active civic volunteer, enjoy cooking, travel, camping, and making music with their children, Joshua and Kathleen.

E-mail: *cmess@drury.edu*, <http://www.drury.edu/info/departments/phil-relg/ess.html>

Fay Sudweeks is a Senior Lecturer in Information Systems at Murdoch University, Australia, a doctoral candidate in Information Technology and teaches Organizational Informatics and Internet Studies. She received a Bachelor degree (Psychology and Sociology) and a Master (Cognitive Science) degree from the University of New South Wales. Her current research interests are social, cultural and economic aspects of computer-mediated communication and CSCW, group dynamics, and e-commerce. She has published six edited books, eight edited proceedings, and thirty papers in journals, books and conference proceedings. She is on the editorial board of the *Journal of Computer-Mediated Communication*, the *Journal of Issues in Humanities Computing*, and the *WebNet Journal*. She has given lectures in Israel, Sweden, Russia, South Africa, Turkey and Germany.

E-mail: *sudweeks@murdoch.edu.au*, <http://www.it.murdoch.edu.au/~sudweeks>

Johannes M. Bauer is an associate professor in the Department of Telecommunication at Michigan State University. He received a Ph.D. in economics from the University of Economics and Business Administration, Vienna, Austria. In 1990 he joined the faculty of Michigan State University as an assistant professor. His research interests are the evolution of competition in telecommunications and energy, the design of optimal regulatory policies, the impact of different legal and institutional regimes on the performance of infrastructure industries, the domestic and international strategies of infrastructure service providers, and policies supporting the deployment of advanced telecommunications technologies. He has published widely on issues of telecommunications and regulatory reform.

Barbara Becker is a philosopher at the Institute for Autonomous Intelligent Systems, National Center for Information Technology (GMD), Germany. Using philosophical and sociological approaches, her areas of research and publication include body and identity, cognitive science (embodied mind), phenomenology and critical theory, virtual identities and embodiment in virtual environments, and virtual communities. She enjoys nature and many kinds of music, art, literature of the twentieth century, and cinema.

Robert J. Fouser is Associate Professor of English and linguistics at Kumamoto Gakuen University in Kumamoto, Japan. His major fields of interest are sociolinguistics, second language acquisition, computer-mediated communication, and Korean cultural studies. He has published numerous papers in these fields in Asia, Europe, and the United States. He writes a weekly column for *The Korea Herald*, a major English-language daily in Seoul, and has written several vernacular papers in Japan and Korea. Besides journalism, his hobbies are photography, mountain climbing, and language learning.

Lorna Heaton is sessional lecturer at the University of Montreal. She has travelled widely in Asia, Europe and North America. She has recently published in the *Journal of Information Technology* (1998: v. 13), *Industry and Higher Education* (1998 v. 12 no.4) and the *Journal of Engineering and Technology Management* (fall 1999), and is completing a book which explores the impact of culture on computer and software design in Scandinavia and Japan.

Susan Herring is Associate Professor of Information Science and Linguistics at Indiana University. One of the first researchers to investigate gender differences in computer-mediated communication, she has also pioneered in the application of linguistic methods to computer-mediated discourse and the study of change in Internet communication patterns over time. She has lectured on CMC in the US, the UK, Canada, Mexico, France, Denmark, Norway, Sweden and Japan, has been interviewed on CNN and the NBC Nightly News, and her work has been written up in Newsweek and the New York Times. Her publications include *Computer-Mediated Communication: Linguistic, Social and Cross-Cultural Perspectives* (John Benjamins, 1996), *Computer-Mediated Conversation* (Oxford University Press, forthcoming), and numerous articles on CMC in books and journals. Currently her research interests center on the impact of emerging CMC technologies on social interaction; the effects of user demographics and context on CMC use; and the applications of CMC research to system design.

E-mail: *herring@indiana.edu*

Soraj Hongladarom is an assistant professor of philosophy at Chulalongkorn University, Bangkok, Thailand. His most recent publications include a Thai-language monograph on *Horizons of Philosophy: Knowledge, Philosophy and Thai Society*, to be published by the Thailand Research Fund, as well as many articles on epistemology, philosophy of language, philosophy of science, and the philosophical and social aspects of Buddhism. Currently he is undertaking a two-year research project, granted by the Thailand Research Fund, on science in Thai culture. He is also organizing a panel on this theme at the 7th International Conference on Thai Studies. Papers from this panel, guest edited by him, will appear in a special issue of ScienceAsia. His hobbies are walking, swimming, and immersing himself in the Internet.

Herbert Hrachovec is an Associate Professor of Philosophy at the Department of Philosophy at Vienna University. He administers a wide range of net-resources of philosophical interest, including the German-language listserv, *register*. He publishes extensively in the area of computer-mediated communications.

Homepage: http://hhobel.phl.univie.ac.at/~herbert.

Kenneth Keniston is Andrew Mellon Professor of Human Development and Director of Projects for the Program in Science, Technology and Society, Massachusetts Institute of Technology.

Carleen F. Maitland is a doctoral candidate in Systems Engineering, Policy Analysis and Management at Delft University of Technology in the Netherlands. She is writing her dissertation on the role of trust in the adoption of electronic commerce by small to medium sized enterprises in both developed and developing countries. In general her research interests include cultural factors and the effects of regulation in the adoption of communication technologies. She is co-author of a paper in the *Journal of Broadcasting and Electronic Media* on V-chip policies in the US and Canada. Prior to her position at Delft, Ms. Maitland was a doctoral student in the Dept. of Telecommunication at Michigan State University. Her hobbies include tennis, rowing, and cycling.

Steve Jones is Professor and Head of the Department of Communication at the University of Illinois - Chicago. He has been Internetworking since 1979, beginning with the PLATO system at the University of Illinois, Urbana-Champaign. He is author of five books, including *Doing Internet Research, CyberSociety* and *Virtual Culture*. A social historian of communication technology, his books have earned him critical acclaim and interviews for stories in newspapers of record and leading news magazines. He has also been interviewed on radio (including National Public Radio) and TV. Jones's interests in technology and policy are also evident in his research into popular music, youth culture and communication. He has published numerous journal articles. He is co-editor of *New Media & Society*, an international journal of research on new media, technology, and culture and edits New Media Cultures, a series of books on culture and technology for Sage Publications.

Additional information can be found at <http://info.comm.uic.edu/jones>.

Nandini Sen is a doctoral student in the Mass Media and Communication program at Temple University, Philadelphia. She has presented papers on telecommunications and computer-mediated technology at national and international conferences and publishes in the field of computer-mediated communication. Her dissertation examines telecommunications policy in Asia. Nandini worked for a multinational advertising agency before returning to academics. Reading, creative writing, gardening, and travel occupy her spare time.

Lucienne Rey is naturally interested in comparing different cultures: her father is from the French part of Switzerland and her mother is from the Italian part, and she grew up in the German-speaking capital of Bern. After her studies in cultural geography she focussed on the question of general perception in the different language groups of Switzerland. She is the author of two books on various ideas of nature and their development in the German, French and Italian part of Switzerland. She currently works for the Education and Development Foundation (Bern), focusing on North-South questions, intercultural activities, and sustainability.

Concetta M. Stewart worked at AT&T for twelve years in international market development, planning, customer and market research, and product management and introduction. She received her Ph.D. in Communication and Information Studies from Rutgers University, and is now a faculty member in the School of Communications and Theater at Temple University, where she teaches courses in telecommunications, policy and organizational communication. She is also a member of the Graduate Faculty in Mass Media and Communications, In addition, she is a Senior Associate for the American Association of Higher Education's TLT Group. She has served on the editorial boards of several journals including *Academy of Management Executive, Journal of Applied Communication Research,* and *Studies in Technological Innovation and Human Resources.*

She lives in Ringoes, New Jersey with her husband Thom Pooley, daughters Sarah and Frances, and three cats. She and Thom enjoy jazz, British sports cars, and travel—her daughters enjoy the travel.

Josef Wehner is a sociologist at the Institute for Autonomous Intelligent Systems, German National Center of Information Technology (GMD). His research focuses on the relationships between the evolution of media and the modernization of society. His current study deals specifically with the loss and recovery of social boundaries in the modern world. His publications include *Ende der Massenkultur? Visionen und Wirklichkeit der neuen elektronischen Medien* (Frankfurt/M.: Campus 1997) with Werner Rammert et al.; *Wissensmaschinen. Soziale Konstruktion eines technischen Mediums. Das Beispiel Expertensysteme* (Frankfurt/M.: Campus 1998); and "Boundary activity"—Zum Verhaltnis von elektronischen Medien und politischen Parteien, in: K. Imhof and Otried Jarren, (Eds.), *Steuerungsprobleme in der Informationsgesellschaft*, (Opladen Westdeutscher Verlag 1999).

Deborah Wheeler teaches at the Henry M. Jackson School of International Studies at the University of Washington, where she is also a fellow at the Center for Internet Studies (see <http://www.cis.washington.edu>). Her Ph.D. in Political Science and Middle Eastern Studies (University of Chicago) reflects her specialization in the study of media in Middle Eastern conflicts. She received a Senior Fulbright post-Doctoral research grant to Kuwait in 1996. In 1992 she was a MacArthur Fellow, sponsored by the Center for the Advanced Study of Peace and International Security at the University of Chicago. She has recently completed a book on the Internet in Kuwait and is presently finishing a book manuscript on media relations in the Palestinian-Israeli conflict. She lives on an island in Puget Sound with her academic spouse, and enjoys camping, gardening, and being outdoors with their three sons.

Sun-Hee ("Sunny") Yoon received her Ph.D. in communication from Oregon State University, and has taught in the Department of Communication, Sogang University (Seoul, Korea). She currently works for the Korea Broadcasting Institute. She has published extensively in the area of CMC, including as a co-editor of *Democratizing Communication: Comparative Perspectives on Information and Power* (Hampton Press) and *Making up a Nation's Mind: South Korean Communication in Transition* (Ablex).

Index